EXPERIMENTAL MUSIC IN SCHOOLS

Towards a New World of Sound

BRIAN DENNIS

Music Department
Oxford University Press
44 Conduit Street, London W1R ODE

SBN: 19 323195 6

First Published 1970
Reprinted 1971

As a supplement to this book, an envelope has been prepared containing separate copies (one each) of Materials 1 – 20. These may be found helpful when the Materials are used in class.

The piece on page 24 from George Self's *New Sounds in Class* is reproduced by kind permission of Universal Edition, London.

Printed in Great Britain by Hollen Street Press Ltd, at Slough

CONTENTS

INTRODUCTION

The health of an art is in danger if those who teach it fall too far behind those who practise it. This book is written to help teachers who would like to introduce truly modern music to their classes. It is designed to help them in the most practical way: by encouraging them actually to make music, contemporary music, in the classroom. This is the way other subjects are kept alive and vital. Consider the influence of contemporary developments in the plastic arts upon the way 'art' as a subject is taught in schools nowadays. Even the most casual enquiry into the way pupils occupy themselves in these lessons can be a most illuminating experience. One finds a concentration on experiment, on self-discovery, not only in the manipulation of shapes and contours but also with regard to texture and the various materials which make this original use of colour and texture possible.

Being continually involved as they are in experimenting with new techniques and images, modern painting and sculpture make use of a considerable variety of materials. A good many of these are synthetic and require new methods of handling. More traditional materials may be manipulated in ways only made possible by technological innovations – photographic projection, amplification and distortion, electric cutters and welders, silk screen printing, and so on. In the classroom much simpler and cheaper methods are chosen as counterparts to these processes. Materials such as cardboard, pages from glossy magazines, pieces of plastic foam and polystyrene might find their way into a collage to provide unusual images and textures. A simple paint spray can produce ethereal effects which would be impossible with the ordinary paint brush. Beautiful designs can be made out of string and wire (although such geometrical constructions might just as easily be the product of a maths lesson).

The development of painting in the twentieth century has had many parallels in music. In particular a concentration on the use of colour, and a strong concern for texture, have been of equal significance in both arts. A special emphasis was laid on tone colour in music by both Mahler and his contemporary Debussy. Since this aspect of their work was a pronounced aesthetic influence on Webern it is not surprising that, however rigidly constructional Webern became with regard to pitch and rhythm, he retained the most intuitive freedom in his treatment of colour. This imaginative use of what Schoenberg called *Klangfarben,* or the melodic continuity of pointillistic strands of colour, reflects a widespread change of emphasis in the field of composition; contrasts of tone-colour have in many ways replaced the harmonic and melodic continuity of previous music. With Schoenberg, Stravinsky, and Charles Ives, dissonance became as acceptable to the ear as triadic consonance, and because of the democratic use of intervals to be found in their works new possibilities of sound and new areas of experience are opened up.

In the work of Edgar Varèse the possibility of incorporating noise (non-pitched sounds) into composition is explored, and large batteries of percussion instruments find their way into his compositions. With the development of electronic music this change of emphasis (colour replacing melody) is almost complete. The problem still remains of how to relate instrumental sounds (i.e. the use of live performers) to more privately generated studio sounds. Stockhausen's *Kontakte* combines the two, and in some of his more recent pieces such as *Microphonies* 1 and 2, electronic devices are used in conjunction with live performers, so that sounds which had hitherto been the creations of the aseptic electronic studio can now be experienced at the moment of generation. This final synthesis of electronic sound and live performance marks the culmination of a process in which purely instrumental and purely electronic music, in the work of this composer, have provided a continuous inspiration for each other. This kind of interaction has been dominant in the work of many present-day composers, particularly in the field of 'textural composition', as it is aptly termed, which is becoming a widespread genre on the continent. In this music, for example the orchestral works of Penderecki and Ligeti, textures and noise qualities have entirely superseded any kind of construction based on melody and harmony. In other words, although this music is *not* electronic music, it is to this that it owes its inspiration, and the sounds produced are very often indistinguishable from electronic sounds.

2

This is only one view of the development of twentieth-century music. Other accounts may be no less valid, but tend to ignore what to my mind has been the most significant aspect of twentieth-century music. *This aspect is colour:* the imaginative use of pure sound qualities, together with more complex manifestations of overall textures and sound patterning.

The mystery and complexity of individual sounds and the experience of these sounds is the most progressive feature of contemporary music. Both the ultra-rational and generally rigid manipulation of harmony and melody, and their apparent opposite — the random deployment of pitches such as one finds in the music of John Cage — are needed to concentrate the listener's attention on the experimental use of sounds and his subsequent experience of that sound. This emphasis is as much a feature of modern music as is the continuous experimenting with colours and materials in the field of painting.

Just as experimental techniques in painting have to be simplified and carefully adapted for the purposes of classwork, so must the elaborate techniques of most post-Webern composers be considerably modified for use in schools. In contemporary music, however, a much simpler approach is already inherent in the music of the American 'school' of John Cage — simplicity at least in the techniques if not in the elaborate theorizing which so often accompanies this music. It is in these techniques that more practical alternatives may be found to the complex serial and schematic contructionalism of Boulez and Stockhausen, without evading the aesthetic implications of the music of such composers. Such practical alternatives are admirably suited for the development of a new approach in educational music. Through a series of carefully worked out processes, Cage developed a means of relinquishing control over his material so that the performer and finally even the listener have an equal hand in creating and creatively experiencing the components of the music.

I do not wish to overstate the case for Cage's theory. It is sufficient here to regard the freedom (the indeterminacy) of Cage's music as a starting point from which ideas can develop stage by stage, in this case in the classroom. The music produced in this way may have as its ultimate objective a broader sound world, the range of ideas implicit in other major areas of present-day music — pointillistic, textural, electronic as well as indeterminate — but it will be concerned in general with the experience of distinctive sound qualities and their associations.

3

The underlying intentions of the various methods described in the book are several:

1. The system of notation and the methods of sound production are simple enough for pupils of all ages and with no prior knowledge of musical notation, to *participate* — the emphasis is always on participation in one form or another.
2. A method is indicated by which the teacher can move towards a training in conventional notation within the confines of the more radical system. Since this is largely through practical work, much of the tedium inherent in the teaching of traditional notation is avoided.
3. The pupils learn to listen and to listen very intently. Their aural perception can be considerably enhanced. An appreciation of all music past and present can follow spontaneously, in a way which again avoids tedium.
4. The ability to improvise creatively as part of an overall pattern, as well as to coordinate with given hand signals and cues, is developed. Rhythmic improvisation is also encouraged.
5. Simple methods — which are often complex in terms of their overall result — are given by which the individual pupil can create his own music. Such methods range from fully composed music for ensemble or solo instruments, to music created on to tape, schematically, spontaneously, or by a mixture of both (not to mention a mixture of live music with tape). Creativity is encouraged to the utmost degree — a creativity whose roots lie in our own environment and in the work of present-day composers.
6. A fascination for other subjects, particularly science and mathematics can now be directed towards music. Experimenting with sound satisfies one of the most fundamental drives in a young person — namely curiosity or the desire to explore. It also satisfies the desire to actively participate in a corporate activity, rather than passively absorb what may be felt too much of an abstraction. The most beautiful piece of music or the most interesting piece of information can easily fail to move a child. What he discovers for himself as a spontaneous by-product of a practical activity more often fires his imagination.
7. Finally, though several more or less mechanical devices are suggested in the book by which music can be generated, I must emphasize that the ear is the only true judge. If a

piece does not sound satisfactory, it is not satisfactory. But let the pupils make their own minds up about this: do not force your own judgement on them.

4

Very few of my contemporaries speak enthusiastically about the music lessons which they had to attend at school. The fact that they were *made* to listen to this or that piece of music postponed in some cases any real appreciation of music for several years. No doubt considerable advances have been made in the last few years, but enforced listening periods and the learning of traditional notation always used to add up to a feeling of gloom and often do so still. My intention is to provide a set of alternative activities so that the music lesson, whatever else it may achieve, *is* associated with a feeling of lively experiment and corporate activity for *all* the pupils involved.

Chapter I

SOUND OUT OF SILENCE

At all moments in one's life sound is continually present even if one is not aware of it. Finding out just how many sounds one can detect at a given moment can be highly instructive. It is this simple basic idea which I have chosen to act as a starting point for this first series of exercises.

Each pupil should be given a sheet of paper. The entire class should then be asked to preserve as complete a silence as possible; each pupil must restrain himself from making any intentional sound whatsoever. Instead the class must listen carefully for any noises which are at all audible during the given time interval; identify the noises and write down everything that can be detected. The period of silence should last about 10 minutes. Sounds can be identified and divided into four categories as follows:

1. Unintentional sounds made by the pupil himself
2. Sounds from inside the classroom
3. Sounds from inside the building as a whole
4. Sounds from outside altogether

A typical result might be:

1. Sound of pencil on the paper
   Sound of own breathing
   Slight brushing of clothing against the chair and desk
   Faint creaking of chair and desk
2. Faint noises from other children – coughing – pencils at work
   Tick of the clock every minute
   Slight creaking of chairs
3. Sound of another class in progress
   Footsteps in the corridor
   Distant hammering
4. Traffic on the road outside. Aeroplane overhead.
   Birds singing – wind – distant shouting

When the period of silence is ended comparisons can be made, marks awarded, etc.

Depending on the age and intelligence of the pupils, the tabulating of sounds can be made more sophisticated, with the pupils noting the approximate time a given sound occurs. (A large stop clock of the kind used in a physics lab may be started from zero at the beginning of the period of silence. An ordinary wall clock is less accurate but can nevertheless provide a feeling of poetic continuity – the period of time not being felt in isolation, but as part of an infinite continuum of sound stretching throughout all time.) The former categories can then be further subdivided into:— sounds which last the entire duration of the silence, sounds which have a very short duration and whose time of occurrence is noted (e.g. the faint sound of a distant bell at 11.44 a.m.) and moderately long sounds for which the time of beginning and ending is noted (e.g. a distant train shunting from 11.51 a.m. to 12.02 p.m.). The fact that a sound is periodic, such as the ticking of a watch, might also be mentioned. The making of such an accurate chart of sounds requires extreme concentration and alertness.

2

As an extension of the former exercise a number of very quiet sounds may now be added to the 'silent' listening period and a new kind of chart made. Depending again on the age and intelligence of the pupils this could include:

1. An identification of the 'deliberate' sounds or at least a description where positive identification is difficult (e.g. high, low, metallic, wooden, regular, irregular, short, continuous, dying away, etc.)
2. Their association with natural, mechanical, or any other sounds.
   The pupil should be asked to describe briefly any kind of subjective image which the sounds might suggest to his imagination.
3. A chart of the deliberate sounds as they relate in time to any other accidental or natural sounds (see the previous exercise) which are also audible during the time period.
4. The timing of the deliberate sounds — when they occur or when they begin and end, depending on whether they are short or continuous. It would also be possible to represent the sounds diagrammatically, for example:

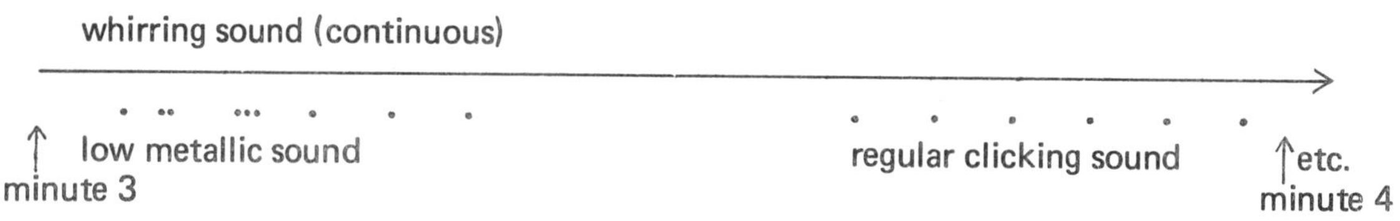

whirring sound (continuous)

————————————————————————→

• •• ••• • • •   • • • • • •

↑ low metallic sound   regular clicking sound   ↑etc.
minute 3   minute 4

If a simple layout of minutes is made before the beginning of the exercise the sounds can be inserted in their appropriate positions as above. For clarity the beginning of each minute can be signalled by the teacher holding up the appropriate number of fingers.

As far as the 'added' sounds themselves are concerned, numerous methods of sound-production can be used. The teacher can if he wishes provide all of them himself, but the pupils respond best if the sound-making objects are concealed. On the other hand a small number of pupils could be instructed to provide the sounds situated, say, at the back of the class or hidden behind a screen. If assistants are employed they must be given simple instructions as to when to make their given sounds. The concealing of instruments is not absolutely necessary, but it helps to concentrate the pupils' attention entirely on what is heard and at the same time turns the exercise into something of a game of the kind that most children respond to. With regard to the selecting of sound sources most conventional instruments would be too loud for this purpose when played in the normal way. A whole host of 'ready-made' instruments could be used instead, like a slowly flicked pocket comb or the tinkling of coins in the pocket. The teacher is at liberty to invent any system of sounds that he wishes and disperse them throughout the period of 'silence' according to any pattern. Here are some more examples of sound sources that might be used:

1. Suspended glass chimes which will generally make soft intermittent sounds throughout, being subject to slight currents of air.
2. Two or three notes of a glockenspiel played with padded beaters or fingers
3. Hands rubbed together
4. Two pencils lightly tapped together
5. Suspended cymbal (also gongs or bells of any kind) tapped lightly with the fingers
6. Fingers or finger nails drawn lightly over the surface of a drum
7. Tissue paper crumpled or a piece of paper slowly torn
8. Any hollow object (not making a definite note) blown over
9. The surface of a glass or any glass object rubbed

etc. etc.

A more substantial list of instruments (for all purposes) is given at the end of Chapter 2. One must remember however that as this is an extension of the extreme quiet of Exercise 1, the deliberate sounds should be scarcely audible. It is suggested that no more than six different additional sounds should be used within the given time period (8 – 12 minutes.) and on the whole these should be widely spaced and contrasted. The form of this exercise has been deliberately left as open as possible so that the individual teacher can decide the degree to which it may be made into an ear test or a game or into an experiment in analysing sound patterns.

3

Although definite pitches may well have found their way into the previous exercises, a precise distinction must now be made between pitched and non-pitched sounds. A simple preliminary exercise lasting two or three minutes can be used to clarify this distinction.

A few deliberate sounds should be made and the pupils should divide them into pitched or non-pitched in the order that they occur. To avoid confusion, semi-pitched instruments like suspended cymbals and other related types should not be used here. Now that the distinction has been made, three chime bars or three glockenspiel plates should be set up. The chosen pitches might be F, B, and E and these should be sounded two or three times before the listening period begins. At the same time they should be described to the class as low, middle, and high (L, M, and H), unless the class has a prior knowledge of notation in which case the letter names of the notes should be sufficient. The pupils must then remember these pitches so that when the listening period begins they are able to differentiate between them when they occur, writing down their pitch or approximate pitch (L, M, and H) in order of sequence. The listening period should be about five minutes and the teacher should disperse the notes slowly but irregularly, with a different number of times each note is repeated, throughout the time period (taking care of course to make a personal note of what he has played).

At the same time the pupil could make a note of the approximate time interval separating each sound. This can best be represented spatially, e.g.:

M   H L   H M   LH L   M H   H

Again unintentional or external noises may also be noted down as well as any non-pitched sounds the teacher might like to add (thus combining both the previous exercises as well). An elaborate series of short ear tests can now be devised by the teacher following the implications of this type of procedure. Up to five or six notes could be used for more advanced pupils. If the designations Low, Middle, and High are retained however, it is very important that the relativity of these terms is made clear. For instance, if as a sequel to the F,B,E exercise suggested above the notes C, F, and D are chosen as Low, Middle, and High respectively, the 'Low' note of the former exercise becomes the 'Middle' note of the new exercise and this must be carefully explained.

After a few attempts at this kind of test have been made, the concept of simultaneous pitches (harmony) can be introduced and with this the concept of relative consonance (or dissonance) may also be introduced. In other words this particular exercise can be used as a basis for conventional ear training – not so conventional perhaps if the teacher, as I hope, retains varying durations (so that time remains relative) and also continues to select a full range of intervals with little or no tonal bias.

4

One definition of noise is that it is unwanted sound. So far all the deliberate sounds made in these exercises have been chosen to merge with all the unintentional environmental noises which have coincided with them in the given area of time. The pupils have been made continually aware of the equal status of these sounds because of the deliberately indeterminate nature of the additional sounds and their lack of organization in time. If a piece of music is played on the gramophone, all other noises can be taken as external. On the other hand a chart can be made of all the external sounds as in the first exercise with the possible addition of an extra category for charting the unwanted noises inherent in the playback mechanism: humming, reverberation, scratches on the record, or a generally noisy surface. This emphasis on sounds which are external to the music will produce as a spontaneous by-product, a concentration on the music itself. The atmosphere of quiet concentration necessary for the former exercises has now established itself as a suitable atmosphere for listening to music. Quiet undramatic music should be selected and the volume of the gramophone should be kept very low. The most conventional of 'class-room music' should be generally avoided; chamber or piano music of such composers as Schubert, Mozart, or Webern would probably be the most suitable. If on the other hand a lighter style of music is preferred, any record by the Modern Jazz Quartet would provide a suitable atmosphere.

Methods of charting prominent features of the music itself, rather than just the external sounds, can now be devised, taking this exercise as a starting point. The general appreciation of music is a very wide topic and is only touched on here. Nevertheless the methods which I would advocate in this connection relate quite closely to the process of 'noting down' which has been one of the most prominent features of this chapter. The graphic representation of pieces of classical music is a recent idea in class appreciation. For instance the Tone-scripts of Kenneth Payne* which present detailed pictures of entire pieces in terms of coloured shapes. Any attempts the pupils themselves might make in this direction would naturally be much simpler, but could use published examples as a model as well as an intriguing means of following the music.

Summary

The concepts of silence and sound, deliberate and intermittent, pitched and non-pitched, functional and non-functional, have been examined in this chapter. Participation has been fostered by accurate listening and noting down what is heard. With the possible exception of the first exercise, each section has contained a variety of expandable ideas which have deliberately lacked detail: the teacher is thus free to develop processes of his own. A variety of different sounds has gradually emerged from a background of 'silence'. This has been the unifying feature of these exercises. In subsequent chapers there will be an increasing degree of specification as well as a demand for more active participation on the part of the pupils. Whether this takes the form of improvisation or the performance of organized pieces, all who take part should continue to listen carefully to the overall pattern of sound that they are making, and remain aware of the all-pervading nature of these sounds and their affinity to 'natural' sounds.

***Published by Tonescript Productions Ltd., 48 High Street, Croydon, Surrey.**

CHAPTER 2

IMPROVISATION

The stage has now been reached when pencil and paper can give way to instruments.* At first the quiet and intimate nature of the previous lessons should be preserved and the practical work considered as an extension of what has gone before. If there are enough instruments to go round every pupil should take part. If the school does not possess sufficient instruments, the pupils can themselves be asked to bring along any sound-making objects to which they have access; in any case such 'instruments' can easily be improvised (see p. 7 , and also the list at the end of this chapter).

The first procedure for improvising is very simple. All the instruments should be of a percussive nature (pitched or non-pitched) and all the sounds made in the given improvisation should be as quiet as possible and well spaced. It is up to the conductor to select sounds spontaneously and to cue them in one by one. Moving from one part of the class to another, he should signal one player at a time with a single short beat. All the sounds should be short (one sound only per beat), but those instruments whose sound takes some time to die away should *not* be damped. Each individual must concentrate on watching the conductor so that he is ready to respond the moment a cue is given to him. The pupils themselves should now be allowed to act as conductor/improviser – the procedure is sufficiently simple. In general each improvisation should now last a somewhat shorter time than the previous exercises.

2

Methods can now be easily devised whereby the class improvises by itself, without needing a conductor. Firstly, each player should be given a number from 1 to 12. (Numbers can be given aurally. Alternatively, numbered cards such as playing cards can be used. The pack should be shuffled and dealt out, usually one card per person. This latter method is partic-

*The teacher may prefer to vary some of the former 'listening' exercises with the more active procedures of this chapter. Our intention is to avoid an 'imposed tedium', and depending on how far the pupils have been absorbed in the listening processes, it might be more beneficial to alternate the listening and performing exercises. The listening exercises can after all be extended and elaborated by the teacher and can also provide a contrast when a number of the practical activities which follow have been attempted.

ularly suitable when more complicated systems of numbering are involved as in later exercises). Every player having received a number, the improvisation should proceed first from left to right, then from right to left. The numbers represent the approximate number of seconds counted by each pupil before following on from the sound which precedes him. To illustrate this here is an example with a class of 25 pupils:

5	10	3	8	1
→	→	→	→	→
4	9	2	7	12
←	←	←	←	←
3	8	1	6	11
→	→	→	→	→
2	7	12	5	10
←	←	←	←	←
1	6	11	4	9
→	→	→	→	→

The arrows on the diagram represent the direction the chain of sounds should take. A more complicated procedure is to start simultaneously from front/left together with third row/left, thereby doubling the number of sounds. A cue from the conductor should help player 1 to pick up from the furthermost player on the right when the time comes.

A second method is to allot each player a much greater number of seconds, say from 10 (the minimum) to 25, preferably with an entirely different number for every pupil involved. The method of assignment is unimportant in this case: the pupils can choose their own numbers if they wish. Each player makes his sound regularly, counting his allotted number of seconds between each repetition. The conductor opens the improvisation with a single cue and the first sound should follow approximately 10 seconds later. The overall pattern is by now quite dense. It might be further enriched by having two or three different numbers to each player (starting lower than 10 if one wishes) so that he either follows these durations in a fixed sequence or permutates them at will.

For example, a player given 7, 18, and 11 may be asked to separate his sounds by 7, 8, and 11 seconds repetitively in that order, or alternatively be asked to permutate the order of the durations in any way he wishes, thus producing a series of durations such as 7, 11, 7, 18, 11, 11, 18, 7, 18, etc. If the latter procedure is adopted pupils should be persuaded to use their time intervals, in whatever order they occur, in approximately equal quantities and not give undue emphasis to the shortest duration.

For the players of instruments whose sound takes some time to die away, numbers can be dispensed with altogether and, instead, the players should be told to make a subsequent sound when the faintest resonance from the previous sound is no longer audible.

A suitable way of ending an improvisation of this kind is to stop each player one at a time proceeding slowly round the class. The best signal for 'stop' is a definite hand movement describing a circle thus: 
whereas the 'start' sign (or the sign for a player to make a single sound as in Part 1 of this chapter) is the usual quick vertical beat thus:

A final method of simple improvisation with single sounds (still as quiet as possible) is designed to give the players an even greater freedom and independence. A number should again be given to each player. If the instrument in question has a long dying-away period the number should not exceed 8. If the sound dies away quite quickly like a triangle the number should be about 12. Players whose instruments do not resonate at all, like a wood block, should be given a number greater than 16. These numbers represent the total number of times each player should make a single sound within the given time period, say 3 minutes. There is no harm in large numbers of players being given the same number; for instance the 8, 12, and 16 already suggested, no other numbers being necessary. The position of a given player's sounds within the duration of the period should be entirely left to his discretion. A clock should be visible to all the performers to give them a good idea of how quickly or slowly they are exhausting their supply of sounds, and in this case a stop-clock is best. Several attempts should be made at this form of improvisation in order to compensate for over-enthusiasm or miscalculation on the part of the players. As with all these methods the teacher is perfectly at liberty to change and adapt, or to invent improvisation systems of his own using these suggested patterns as working examples. Better still, the children themselves should be encouraged to devise such systems. If playing cards are used, the pupils could be encouraged to invent meanings for the different suits, the colour cards, etc. In the final method of this chapter, it is the children themselves who determine the shape of the improvisation. Various different shapes should be tried and the effect of each should be discussed; there is plenty of room here for collective decision-making. In the previous methods, where counting is involved a child's idea of time is not likely to be accurate. This is not important. The idea that the given numbers must represent seconds may be modified if one wishes. Some pupils can be asked to count very quickly and others slowly. The actual speed can be determined by any number of subjective criteria e.g. 'count at the speed you think your heart is beating', 'count at the rate you are breathing', and so on.

3

Up to now only single point sounds have been used. A new cue sign should now be introduced to signify a continuous sound – a roll, trill or tremolo, depending on the type of instrument used. A variety of signs could be used for this purpose. The sign I invariably use myself is that of an extended palm like a policeman's halt sign. This means that when the sign is given to an individual player he must make a continuous roll, trill or tremolo until the conductor returns to him later and signals him to stop. The conductor has enormous scope here to improvise a kaleidoscopic pattern of overlapping tremolos, changing the component colours at will.

Before achieving a system of complete independence for each player (as with the point sounds of part 1), it is best to start by dividing the class into several groups with the members of each group reacting together. The groups may be divided at random if the instruments have been allotted in a random fashion. On the other hand instruments could be distributed to make deliberate colour groups.

A simple division might be as follows:

A	B	C	D
skin instruments	wood and glass non-pitched instruments	pitched instruments	non-pitched metal instruments

Or one might have a more sophisticated grouping of mixed colours:

A	B	C	D
low skin and glass instruments	metal (pitched) & wood (non-pitched)	wood and glass (pitched) & metal (non-pitched)	high metal (non-pitched) & high skin

The series of possible combinations of this kind is almost unlimited, and the children should be encouraged to make decisions of their own in this matter.

As far as pitched instruments are concerned, it is well worth while determining what pitches are to be played according to some simple system such as the building up of chromatic clusters. If for instance there are two large glockenspiels, two small glockenspiels, one xylophone, and a set of tuned glasses, the notes to be played (to the exclusion of all others) could be arranged as follows:

large glockenspiels 

xylophone 

small glocks 8ve...... glasses 8ve

In general it is preferable for the pitched instruments to avoid duplicating the same pitch or pitches. If there are more than six instruments playing, single-note tremolos can be substituted for two-note trills where appropriate. Specifying pitches gives those with a rudimentary knowledge of notation a chance to make use of it.

All rolls should still be as quiet as possible. At this stage however, signs implying crescendo and diminuendo can be introduced to provide contrasts of balance and to introduce, for the first time, a sense of movement. The conventional signals for crescendo and diminuendo are best: up-turned palm moving gradually upwards for crescendo, down-turned moving downwards for diminuendo.

When the possibilities of controlled groups producing blocks of tremolos are felt to be exhausted, each player can react to the tremolo sign individually. This will open up a new range of sound patterns. The introduction of crescendos and diminuendos provides the first break with the extreme quiet of the former exercises. This effect should be used very sparingly at first and should be preceded by several improvisations in which the rolls are scarcely audible.

As in the improvisations devoted to point sounds, the control of the conductor may be relinquished in favour of self-contained improvisations. These can be achieved either by retaining the group formations or with each player working individually. A specific time or times should be given to each group or player (a precise timing if a stop-clock is used — say 1'12" — or an approximate time on the ordinary clock say — 11.16 a.m). At the given time the group or individual should start to make a continuous tremolo. The approximate duration of the tremolo should also be given: the group or individual should terminate the roll after counting, say, 20 seconds if this was the number given. A variety of durations should be given to the players, ranging from about 15 to 30 seconds. The total duration of such an improvisation should be from 3 to 5 minutes. Given certain limits the pupils can make their own parts.

A chain process can also be set up so that the players within a given group follow on from one another — so that each player takes over when the player immediately preceding him comes to the end of his roll. Again a series of durations must be distributed amongst the players in a deliberately random fashion. If the groups run in rows from left to right the durations can be allotted from front to back. Here is an example with twenty pupils (the numbers being the durations in seconds given to each of the players):

group						
D	24	36	21	33	18	←
C	21	33	18	30	15	→
B	18	30	15	27	39	←
A	15	27	39	24	36	→

As indicated by the arrows, the chains can move in opposite directions. There is also no need for them all to start together; each group can be given a separate cue to start. The teacher can wait about 10 seconds between entries or alternatively the first player of each group can estimate a 10 second delay before following on from the group in front. If the latter method is chosen each group leader can make a signal as he begins to play so that the opening player of the next group can begin to count his 10-second delay more easily.

As a slight variation each player (or alternate player) could be asked to make a slight crescendo and diminuendo towards the centre of the roll.

When improvisations with points sounds and rolls, taken separately, have been successfully accomplished, attempts should be made to mix these two fundamental elements.

The players should start by reacting together in groups, and work towards complete independence. They must now be very alert as to which sign to react to next. Once a signal for a tremolo has been given, the player must remember to continue playing until he is subsequently 'brought off'. The single point-sounds (i.e. one sound per beat) can now be played a little louder to make a greater contrast with the tremolos, the job of the conductor/improvisor can be given to a number of pupils in turn, the conducting techniques still being quite simple. If the class is large a division into groups is best retained for this purpose.

Self-contained systems without a conductor can now be devised which employ a mixture of rolls and point-sounds.

Here are a few examples with 20 pupils divided into four groups

group D (points)	4	8	2	6	12	←
C (rolls)	20	30	15	25	35	→
B (points)	2	6	12	4	8	←
A (rolls)	15	25	35	20	30	→

For the groups playing *rolls,* the numbers represent the timing of each roll, with the next player following on as before. The numbers have a similar significance for the point players: each represents mental count before following on from the previous player in the group. Seeing that a given row of points will finish well in advance of the tremolo groups, the players (groups B and D in this instance) should continue back and forth until they receive a final 'off' from the teacher, or, until no more rolls are audible from either or both of the roll groups.

The next example continues with the same system for the roll groups, but with an individual counting procedure for the point groups. The numbers allocated to or chosen by each of the point players represent the approximate time in seconds between each sound made by the individual concerned, and may be taken continuously in that order or permutated ad lib.

D (rolls)	30	20	35	25	15	←
C (points)	25, 15, 5	17, 7, 27	9, 29, 19	31, 21, 11	23, 13, 33	
B (rolls)	35	25	15	30	20	→
A (points)	16, 6, 26	28, 18, 8	10, 30, 20	22, 12, 32	14, 34, 24	

Finally, here is an example using a stop-clock in which all the players make an individual roll *as well as* a number of point-sounds. Each player is given the exact time when he must make a roll (this is the first number); the second number in brackets gives the duration of that roll and the third number gives the number of point sounds which must be made before or after the roll. Some players could be instructed to make *all* their point-sounds before their roll, some all their points after. This would make for an interesting variety of effects.

1'.0"(15)10	1'.55"(18)8	2'.50"(21)18	3'.40"(24)16	4'.30"(27)14
0'.45"(30)12	1'.40"(33)10	2'.25"(36)8	3'.35"(15)18	4'.15"(18)16
0'.40"(21)14	1'.25"(24)12	2'.5"(27)10	3'.20"(30)8	4'.10"(33)18
0'.20"(36)16	1'.10"(15)14	2'.0"(18)12	3'.10"(21)10	3'.55"(24)8

In all these suggested patterns, the teacher may distribute crescendos and diminuendos to some or all of the roll players. As far as the single sounds are concerned, the players should be told to make a sound slightly louder than the general dynamic level at the moment of playing.

The pupils should at all times be encouraged to devise patterns or to formulate any ideas of their own with regard to these fundamental sound processes.

4

In order to produce an even greater variety of sound-patterning, another, final, signal should now be introduced. For this I use a clenched fist. This sign can mean one of two things — if a pupil is playing a pitched instrument consisting of at least three notes, the first sign means 'improvise freely on these notes' (or on all available notes). Players possessing a full chromatic range can in certain instances be limited to three or four previously determined pitches.

For players of non-pitched instruments the upraised fist means 'play a specific *rhythm*'. Again a variety of rhythms must be allocated or chosen by each of the players, just as pitches are allocated to the pitched-percussion players. Rhythms may be distributed with a minimum of trouble in a variety of ways. A numerical system such as the following is probably the simplest. A single number between 3 and 10 is given to each player. If a player is given, for example, the number 5, he plays the following rhythm:

..... , ... , ... , ... , ... etc.

In other words after playing a group of 5 quick beats a short pause is made before the next group. The length of the pause can be varied ad lib. (say, between 1 and 3 seconds approximately)

For variety two or three figures can be given to each player. If for example these are 2, 7, and 5 the result should be as follows:—

.. , ... , ... , .. , ... , ... , .. , ... , ... etc.

More advanced pupils could be asked to permutate the given numbers ad lib, in which case (with 2, 5, & 7) the result might be something as follows:—

.. , ... , ... , ... , .. , ... , .. , .. , ... etc.

When a player of (for example) a wood-block is given a signal to make a *roll,* he should produce a tremolo as fast as he is capable without it becoming uneven or louder than required. His speed, when he is given the *rhythm* sign, should be at least half this speed. A player with a drum should play his rhythm with one hand only, again at about half the fastest speed he is capable with that hand *on its own.*

When several players make this type of rhythm at the same time, many interesting rhythmic effects are produced spontaneously after a few moments have elapsed. Players sometimes fall into rhythm with each other, but if their pauses are carefully varied they will begin to move against each other again. It can be instructive for the teacher to suggest on a given occasion that the players should deliberately strive to produce the *same* rate of pulse and on another occasion to deliberately produce a different rate of pulse. As far as the pitched instruments are concerned, here are a few suggested groups of pitches which can be allocated to the players and on which they should improvise to the exclusion of all other notes.

Using the principle of non-repetition, groups of pitches can also be decided by the individuals concerned, if they so wish.

Improvising with rhythms should begin at once, each pupil receiving an independent signal. (Groups of instruments producing this kind of rhythm at the same time make for too great a density of sound). As with the tremolo signal, a rhythm should remain continuous until the pupil concerned is brought 'off'.

The best way of introducing 'rhythms' is to mix them almost immediately with rolls; there is sufficient difference between an extended palm and an upraised fist for this not to lead to indecision amongst the players. As an alternative, only a small number should be asked to demonstrate the playing of rhythms at any one time. If the remainder of the class is asked to listen carefully to the result, there should be ample to interest them in the way the component parts interact with each other.

All rhythms should be made fairly quietly, although the crescendo and diminuendo signals can be used in the same way as with tremolo patterns. It is always intriguing to highlight an individual rhythm (or roll for that matter) by having all the remaining pupils playing very quietly and then indicating an individual crescendo so that one player suddenly becomes a soloist.

To ensure a greater variety of speeds amongst the individual rhythmic patterns, general indications of tempo can be given separately to each player i.e. very fast, quite fast, moderately slow, and slow. These should be given in relation to the degree of resonance of the instrument in question. Gongs, large drums, and cymbals should be played slowly or very slowly; claves, small drums, marraccas, etc. should play quickly. It can be intriguing, however, to have the short 'dead' sounds of the latter group playing slowly and interesting effects can also be achieved by having resonant instruments playing quickly. The conductor should feel at liberty to experiment with tempo indications as much as he wishes.

For very slow rhythms a different system to the one outlined above might be used — a system which is in fact rather similar to one used in connection with some of the previous non-conducted improvisations. Two or three numbers should be given to each player, say the bass drum player, and instead of these representing the number of beats in a group of fast beats, they would represent an approximate number of seconds *between* each beat. For instance, if the given numbers are 7,1, and 3, the player should make his rhythm (when signalled) as follows:—

7″ —— 1″ —— 3″ —— 7″ —— 1″ —— 3″ —— etc.

If there are enough instruments for some pupils to play two or three instruments at the same time, interesting ostinato patterns can be made when the rhythm sign is given. If for instance a player has a small drum, a cow-bell, and a wood block, patterns such as this can be devised (the order in which the instruments are struck can be changed or even freely permutated in each of these examples):

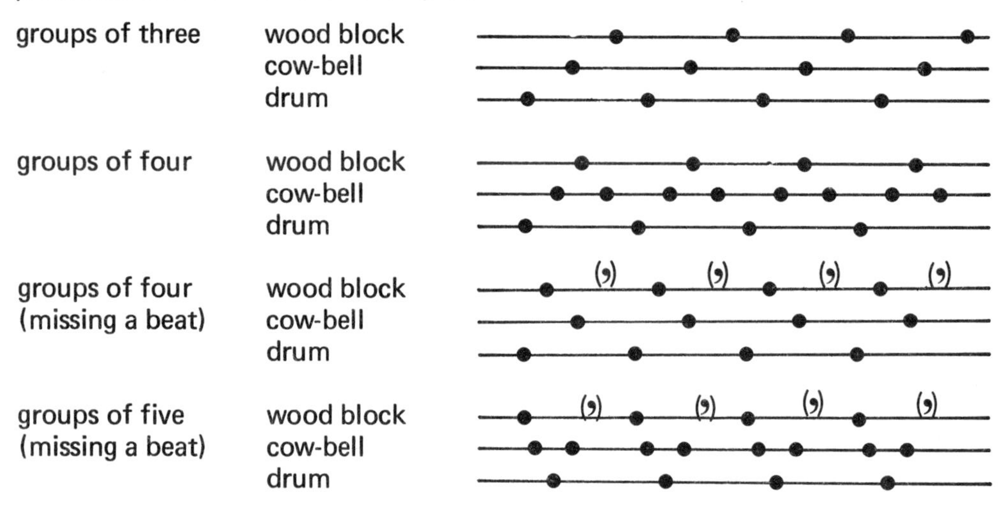

and so on.

All these examples can of course be represented in standard notation:

If the pupils have received a basic tuition in rhythmic values, groups of durations expressed in conventional notation can take the place of the simple numerical system if desired. This stage in improvising is in fact an opportunity for time values to be taught and practiced. Any of the following time patterns can be used and distributed amongst the players:

One of my former suggestions was that the players should try to preserve the *same* pulse. This is probably the best way to start off, if conventional rhythmic patterns are used. The players should play together at the same speed until the rhythms are firmly mastered. When this has been achieved tempos can be varied and rhythms displaced as in the previous systems.

Finally here is an example of a non-conducted improvisation using a mixture of rolls and rhythms in the 'follow on' pattern. The same principles apply to rhythms as to tremolos (i.e. a player should make a roll or rhythm for the given number of seconds and the next player follows on after the first has finished). The rows should start about 10 seconds after each other and continue back and forth until 'brought off'.

group					
D	24	36	15	27	24
	←				
C	21	33	18	30	21
					→
B	18	30	21	33	18
	←				
A	15	27	24	36	15
					→

In this case the teacher, or the pupils, might like to decide which rows play rolls and which rhythms. On the other hand a system of rolls alternating with rhythms within each row might be preferred. Several systems could be experimented with.

By dividing the class into three or even six groups, self-contained improvisations can be devised, utilizing all three types of sounds (points, rolls, and rhythms). The more carefully such systems are determined and the more elaborate they become, the more they resemble the fixed compositions which are the subject of the next chapter.

5

When all the procedures of improvising which are demonstrated in this chapter have been attempted and practised several times, the pupils – and for that matter the teacher – should have become very adept in reacting to and manipulating the various hand signals. Spontaneous improvisations using all three of the basic sounds can become more and more fluent, and at the same time more musically meaningful. The appreciation of scattered sounds, very quiet and random, for their own suggestive values has given way to a far more complex interplay of textures. In many cases, particularly in the more indeterminate improvisations (the self-contained improvisations*), a sense of order has been scarcely apparent. The sounds have tended to be static and uniform, albeit rich in colour. The time has now come to select from this enormous variety of material specific images and to channel them into determined structures. Before embarking on this, however, here is a list of suggested instruments. This gives an idea of the sort of sounds which can be used either for the improvisations in the previous two chapters or throughout the remainder of the book.

*One could quite justifiably disagree with the use of the word 'improvisation' for this type. Some of them are in fact quite strictly determined. The *total effect* is however more improvisatory than in the conducted category, which is why I have continued to use the word.

A SHORT INVENTORY OF POSSIBLE INSTRUMENTS

Non-pitched percussion — Wood

1. Claves or equivalent objects (any two pieces of wood can be beaten together to produce the effect of claves, even two rulers).
2. Castanets. Ordinary hand castanets are suitable for producing point sounds and rhythms. Castanets on a stick produce very good rolls but are generally more clumsy for other purposes.
3. Wood blocks and Chinese blocks. The former are very easily made but should be quite resonant. Sets of Chinese blocks were very popular thirty or forty years ago but are nowadays more difficult to come by except in second-hand shops.
4. Marraccas. Another standard instrument which is easy to make; for instance an empty plastic detergent bottle can be filled with dry peas. Several pairs of contrasting tone should be acquired if possible.
5. Rattles of all kinds (loud and soft). Football rattles might be a little too loud for this purpose. An assortment of babies' rattles can be highly effective providing no one is ashamed to use them.
6. Wooden drums. African slit drums are the ideal but are almost impossible to obtain. A highly industrious pupil could make one by hollowing out a log. Simpler equivalents can be made out of sturdy wooden boxes of varying sizes.
7. Wobble board. This is a highly effective instrument and extremely easy to make. All that is required is a piece of hard-board; several pieces of different sizes produces a set. It is sometimes difficult to maintain a consistent pulse by shaking, but each wobble board will become more flexible the more it is played. Smaller ones should be held in both hands. Larger ones should have one end resting on the ground but firmly placed so that it does not slip.
8. Wood chimes. This is another instrument which can easily be made if not purchased. Approximately 16 pieces of hard wood should be suspended by 2 or 3 inches of string from a cross bar, thus:—

If the chimes are agitated, they should produce a series of clicking sounds which gradually die away. Such an effect is very beautiful but is generally too imprecise to produce even rolls or rhythms. Point sounds can be made by rapidly crushing the chimes together in both hands and then holding them firmly to prevent subsequent clicks.

Non-pitched percussion — Skin instruments

NB. All drums, however large, can be played with a variety of different sticks: side-drum sticks, brushes, timpani sticks, rubber sticks, etc.

1. Hand drums. Bongos, African and Indian drums etc.
2. Large drums. Bass drum, timpani, tom-toms, etc.
3. Small drums. Side drums, tambours (Orff model), etc.
4. Tambourines

(As well as varying the type of beater used, the area on which the drum is struck should also be continually changed. Even the wooden or metal parts of, say, the bass drum should be struck almost as regularly as the skin surface).

Metal instruments

1. Triangles, or any kind of high pitched metal object which produces a similar sound.
2. Indian bells (sometimes called Chinese or antique cymbals). Again any similar sound producing object may be used, such as steel eggcups or suspended metal bars, provided the semi-pitched characteristic of Indian bells is apparent.
3. Sleigh bells, bell clusters, jingles, etc. A shaken bunch of keys provides a higher pitched version of this sort of sound. Any equivalent cluster of high-pitched metal objects can in fact be used.
4. Cymbals of all sizes from large suspended cymbals to small hand cymbals. It is surprising the enormous number of sounds which can be produced from these instruments by varying the type of beater (thin wooden sticks, triangle beaters, etc.) as well as the striking area of the cymbal (the extreme edge and central hub in particular). A suspended cymbal bowed along its edge with an old violin bow creates a marvellous series of sonorities.
5. Gongs of all sizes. Similar sounding objects such as pans, pan lids, and tea trays can also be used to great effect.
6. Cow-bells (authentic cow-bells are hard to come by and jazz cow-bells have nothing like the same effect). Approximately cow-bell-like sounds can be produced from hollow metal objects such as empty tins. These should be as sturdy as possible.
7. Oil drums. An empty oil drum is a ready-made instrument capable of an enormous variety of sounds. Again, sticks and striking areas should be continually varied. Small hard objects dropped inside the oil drum (lead shot for example) will produce a very resonant sound. Oil drums can also be tuned as they are in West Indian steel bands. This is done by indenting the top surface with a hammer and blunt spike. If this is done carefully a wide range of pitches can be played on the same drum.
8. Brake drums and hub caps. Broken down machinery of almost any kind can provide innumerable percussion instruments. Two of the most accessible as well as the most potentially musical parts of a car are the brake drums and the hub caps. The latter usually has a shallow high pitched ring and the former a more resonant gong-like sound.

Glass percussion instruments

1. Bottles. Almost any kind of bottle may be struck with a hard stick to provide a good sound, one which is in many ways superior to the more expensive wood block.
2. Glass chimes. The principle is the same here as with the wooden chimes, although the pieces of glass are more often circular. This instrument can be easily obtained from shops

selling decorative household goods. Chimes made from pieces of stone are also easy to come by, although the 'sound-cluster' of this instrument dies away much more rapidly.
3. Glass marraccas and rattles. A certain degree of ingenuity can produce very attractive instruments of this kind. Sealed jars containing pieces of glass or porcelain have a very different timbre from wooden marraccas.

Other materials

1. Plant pots. A surprising range and variety of sound can be achieved here and it is often possible to tune them quite accurately.
2. Articles of crockery can be struck to provide many more interesting sounds but *beware breakages!*
3. Mechanical objects. Any clockwork or battery-powered motor can produce an interesting whirring or buzzing sound.
4. Wobble boards do not have to be made exclusively from wood. Pieces of plastic, card etc. can provide a miniature variety.

Note. So as not extend this list out of all proportion, the four examples in this section represent only a minute selection of the enormous number of instruments one could devise.

Pitched Percussion

1. Glockenspiels of all sizes and ranges.
2. Chime bars of all sizes and ranges.
3. Xylophones of all sizes and ranges.
4. Wooden bars of all sizes and ranges.
5. Tubular bells and hand bells (preferably suspended).
6. A range of drinking glasses can be tuned by adding water.
7. Glass bowls (for low resonance) or any other glass containers tuned as above.
8. Auto-harps, chordal dulcimers etc. Although these instruments are usually plucked, an interesting series of sounds can be obtained by striking with various kinds of beater.
9. Old piano. As with number 8, a variety of sticks on the inside strings can produce a vast number of sonorities. If the pedal is held down, resonance is enormously increased. Even the body and metal parts of the piano's inside can be used to great effect as non-pitched percussion.

Note. Many of the instruments mentioned in the previous sections have potentially been pitched or near-pitched instruments (i.e. brake drums, cow-bells, plant pots, etc.). I have even mentioned timpani in the context of skin instruments. Orff tambours and tom-toms can also be tuned to some extent. But these instruments have been kept distinct, because I do not wish to have them function as 'pitched instruments' in the next two chapters. They may however, be used freely in other parts.

I am not proposing to make a list of string or wind instruments. Many of the examples in the next chapter can easily incorporate such instruments, but I do feel that these instruments should by and large play the sort of music for which they were conceived, and my own concern for pointillistic colours and improvisatory rhythms, rolls etc. was not developed with traditional orchestral instruments in mind.

CHAPTER 3

PIECES FOR THE CLASSROOM

In the previous chapter, where the conductor improvised with collective groups of instruments, the coordination and the quick reaction to a given signal became all-important. Both the partially and fully determined pieces set out in this chapter are designed to produce an increasing degree of collective response. First however we must deal with notation.

One of the problems in modern music which confronts the performer, and to some degree even the listener, is the number of different systems of notation which are used by composers at the present time. This unprecedented variety has been produced by the degree of experiment which is such a prominant, even fundamental, attribute of modern music. Pitch, duration, density, dynamics and the unlimited ways in which sound can be produced by different instruments — all have produced innovations in the field of notation. Symbols have been adapted, and even specifically invented, to embody the procedures around which the pieces are constructed. In the field of school music however, experimental composers are at present sufficiently few for a more or less standard notation to have been adopted. In England, George Self, David Bedford, and myself have, by personal consultation and being continually aware of each others' work, attempted to standardise such a system. Derived to some extent from the work of such widely different composers as Stockhausen, Berio, Cage and Cardew, this is by and large the simplest and most consistent set of symbols which we could devise for this purpose. George Self, the originator of this approach to class music, was very largely responsible for devising this system himself, although the notation and symbols set out below do suggest one or two additions and alternatives to his approach.

●= a short sound without resonance
●⌢= a sound whose natural resonance is allowed to die away
●˥ = a resonant sound damped about one second after striking (if the sign ● is given to a resonant instrument, say a cymbal, the instrument should be *held when struck* to prevent resonance)
/\/\/\/\/\/\ = a roll, tremolo or trill (the more conventional trill sign — ∿∿∿∿ can be used for pitched instruments)
\/\/\ = a short roll (2 or 3 seconds)
⟋ or ⟍ = a glissando (upwards or downwards)
●—7″—●= approximate timing between attacks
2″ = a pause of 2″ duration
,

('') = similar to above but represents missing beats in a chain of pulses

.... 2'' ... 1'' .... 4'' = a rhythm similar to the improvised rhythms of the previous chapter.

= a two-note chord on a pitched instrument allowed to die away

= an improvised group of notes (ascending) on a pitched instrument.

Conventional dynamics are used although the size of a note sometimes denotes its loudness (as in Stockhausen's *Zyklus*).

● = ff ● = mf ● = mp • = pp

beaters: = hard sticks = soft sticks

||:...,..,....:|| or ...,..,....etc. = continue to repeat as far as indicated (in relation to the system of beats — see below)

Beats

The simplest system by which a piece can be controlled is for the conductor to make a specific number of beats. These beats should be carefully counted by each of the players. The time interval between each beat must be estimated by the conductor himself. If, say, twelve beats are necessary for a short composition, the layout might be something like this:—

(1) ↓ 8'' (2) ↓ 2'' (3) ↓ 9'' (4) ↓ 5'' (5) ↓ 4'' (6) ↓ 2'' (7) ↓ 3'' (8) ↓ 6'' (9) ↓ 5''
(10) ↓ 4'' (11) ↓ 12'' (12) ↓

To give the conductor greater freedom the beat system is sometimes more generalized with a smaller number of given durations and with the conductor calculating the beats within these durations according to their relative spacing, e.g.:

30'': (1) ↓ (2) ↓ (3) ↓ (4) ↓ (5) ↓ (6) ↓
40'': (7) ↓ (8) ↓ (9) ↓ (10) ↓ (11) ↓ (12) ↓

Most pupils should be able to follow a piece of between 25 and 30 beats without getting lost. Beyond this number, depending on the complexity and variety of what they are expected to play, a system of sections might be adopted. Each section would then consist of a specific number of beats (say 12 beats per section) and so as to define the beginning of each section clearly, the conductor can hold up a large printed letter (A, B, C, D, etc.) which would then signify the beginning of sections A, B, C, D, etc. To avoid confusion the card should be held in one hand and then a moment later, when the card has clearly been seen by all the players, the first beat of the next section should be given with the right hand with the letter card still visible. My own *Stars* is an example of a piece divided into sections (four sections of 12 beats) but lettered cards are not necessary here because each section is followed by a short piano solo and a simple cue is all that is necessary to signify the beginning of a new section.

A series of twenty-two pieces all using a system of beats is to be found in George Self's *New Sounds in Class* (Universal Edition). Here as an example is piece No. 12 consisting of 20 beats:

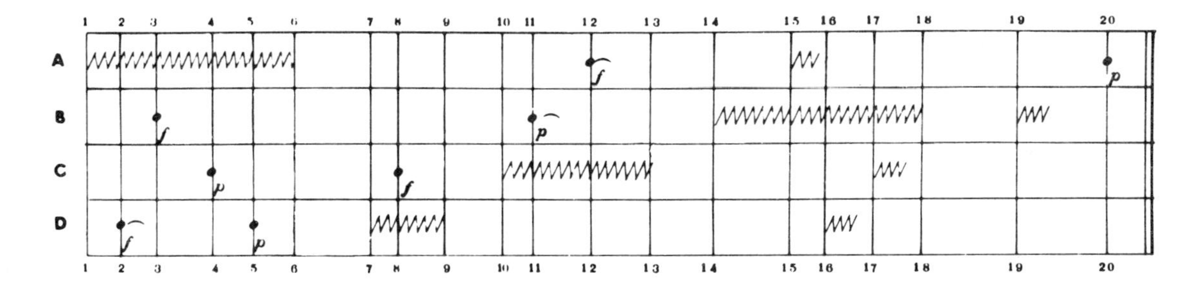

The division of the players into groups is up to the discretion of the conductor, although Self makes several recommendations with regard to the mixing or separating of different tone colours. The general timing of the piece is also up to the conductor and may depend on how many instruments are playing and the range of each.

The group of pieces which now follow have a dual purpose. They can be used simply as classroom pieces of progressive complexity; or they can be used at a later stage as blocks of material in conjunction with the making of electronic music (see Chapter 5). Each piece will be labelled Material 1, Material 2 etc. and will be referred to as such in Chapter 5. In pieces where the instrumentation is precisely indicated any number of instruments approximating to the given tone-colour may be used for each part, but in Chapter 5 the given instrumentation must be used more precisely. It will also be noticed that these pieces are generally more symmetric in design than Self's, as well as relying to a large extent on the repetition of equal durations. These techniques have become a predominant feature of my music as a whole and, although a good deal of choice is still inherent in the way these pieces can be 'orchestrated' and subsequently made into a large number of possible electronic pieces, very little indeterminacy is apparent with regard to their individual structures. My intention has been to create a series of specific musical images and audible sound patterns.

Material No. 1 Duration 1'12"

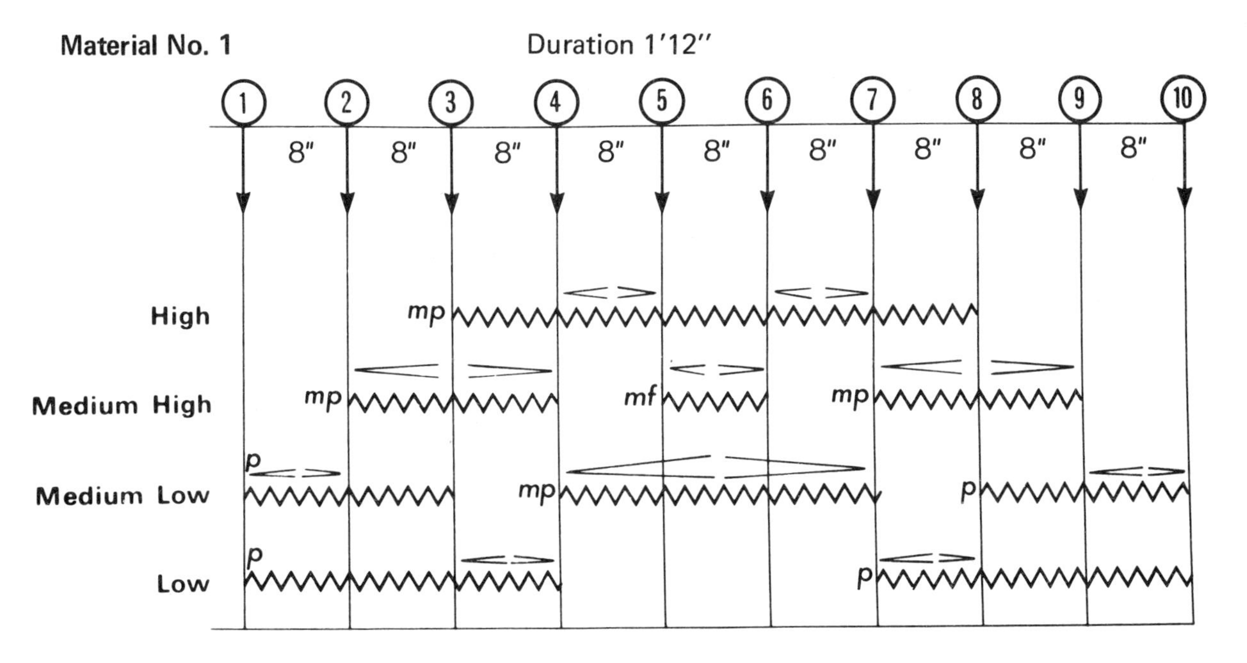

This first piece consists entirely of rolls or trills. The relative pitches of the instruments should determine what part they should play. Crescendos and diminuendos can be indicated by subsidiary signs (see Chapter 2 part 3), otherwise the system of beats is exactly as described earlier. Instruments of all kinds can be used as long as there is a fairly even number of instruments for each part.

Material No. 2 Duration 2'0"

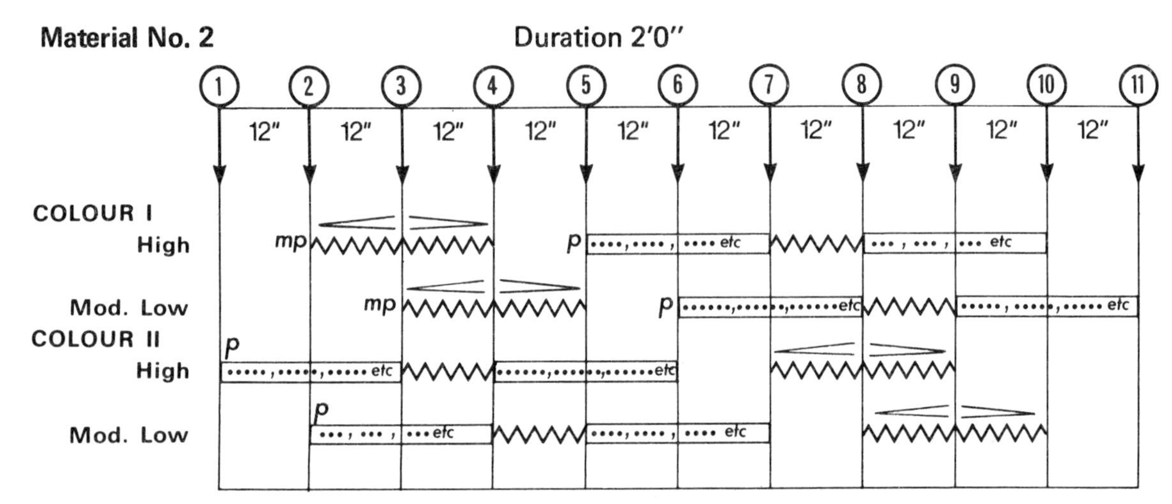

Material No. 2 introduces boxes of rhythms (see Chapter 2 part 4 on rhythmic patterns). Each rhythm, which should never be played too quickly, is repeated over and over for the duration of the box. If possible only two basic colours should be used (say skin and metal); these should then be divided into high and low to produce four parts altogether.

Material No. 3 Duration 2'0"

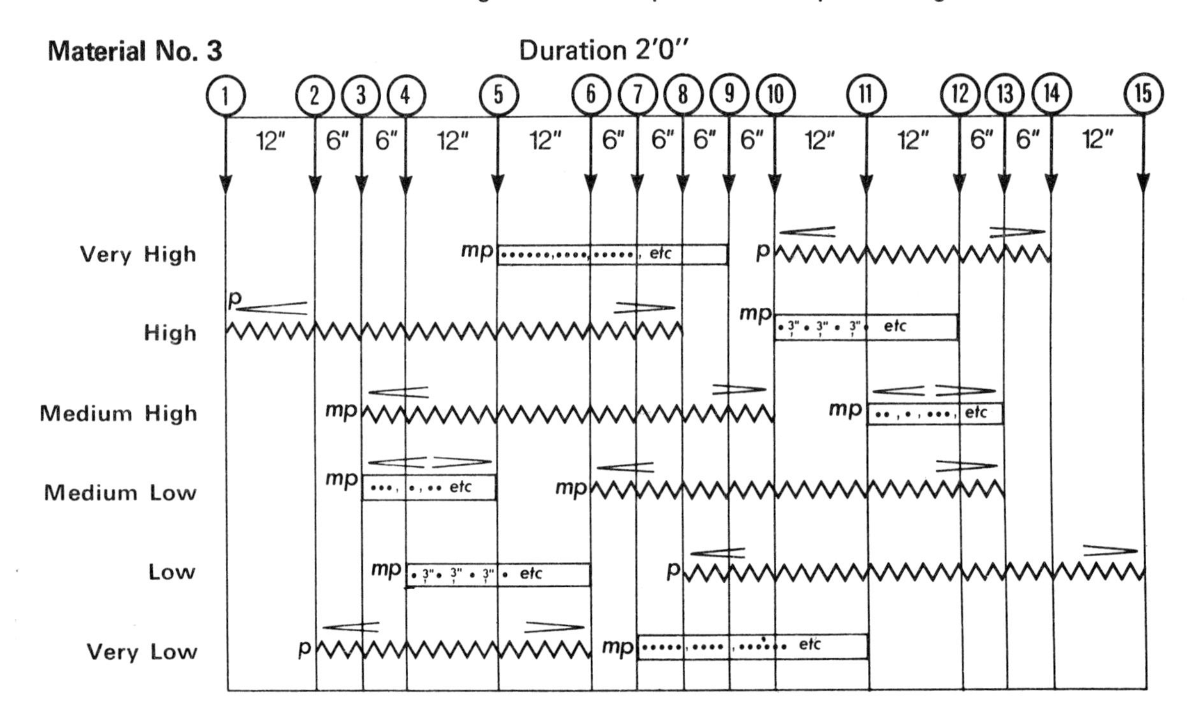

Here is a more elaborate combination of rolls and rhythms in six parts. Division is made according to relative pitch although different colours can be used for the extra clarification of each part. Note that two of the rhythm boxes contain slow, regular beats (3 secs. between each).

Material No. 4 Duration 1'0"

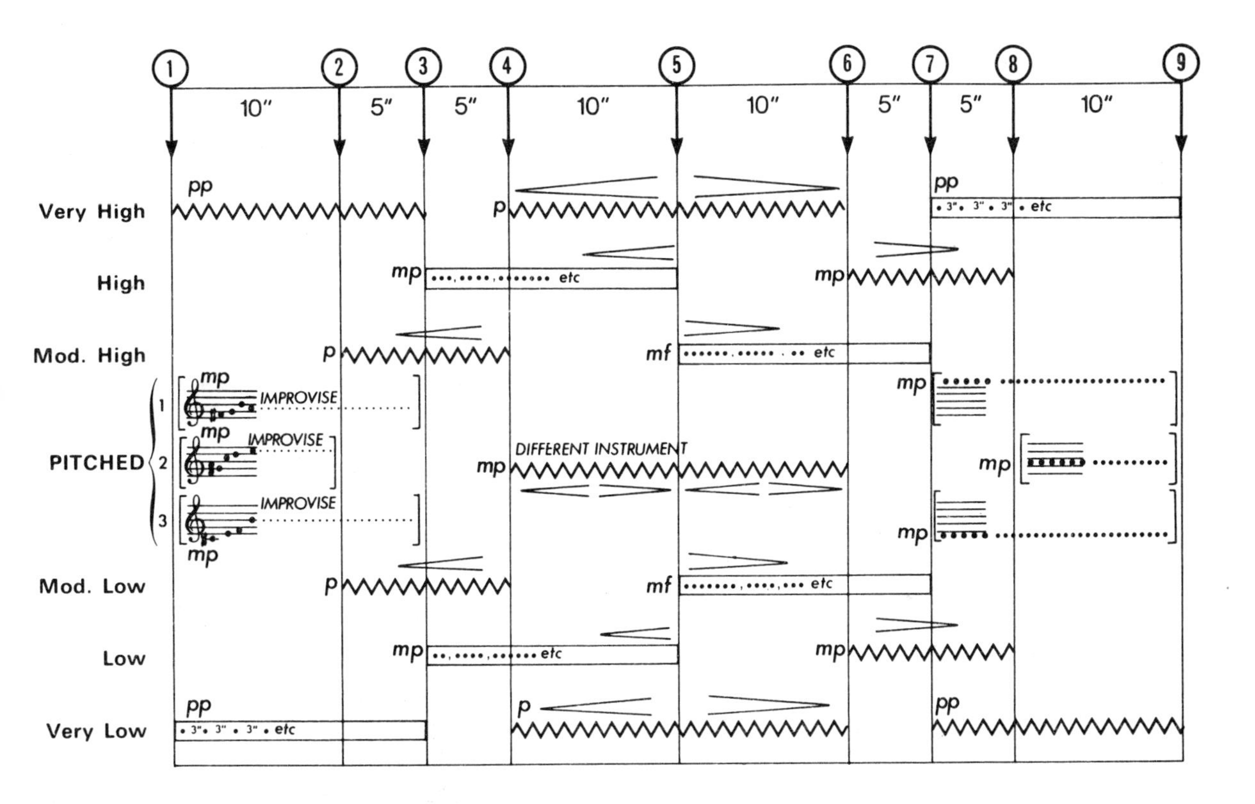

This piece gives a prominent part to pitched instruments. (N.B. There is no reason why pitched instruments, used very freely, should not have been included in Materials 1–3.) Improvisation on pitched instruments was described in Chapter 2 part 4. If one wishes, a regular beat (moderately fast) can be maintained throughout the piece (i.e. the players can be asked to *keep* rhythm or alternatively be asked to play against each other). There are nine parts – 3 high parts (each of different colour if possible) – 3 parts for pitched instruments and 3 low parts (again of different colours if possible). Players of part 5 should be given additional instruments, preferably of a strongly contrasted tone colour, which play between beats 4 and 6.

Material No. 5 Duration 2′0″

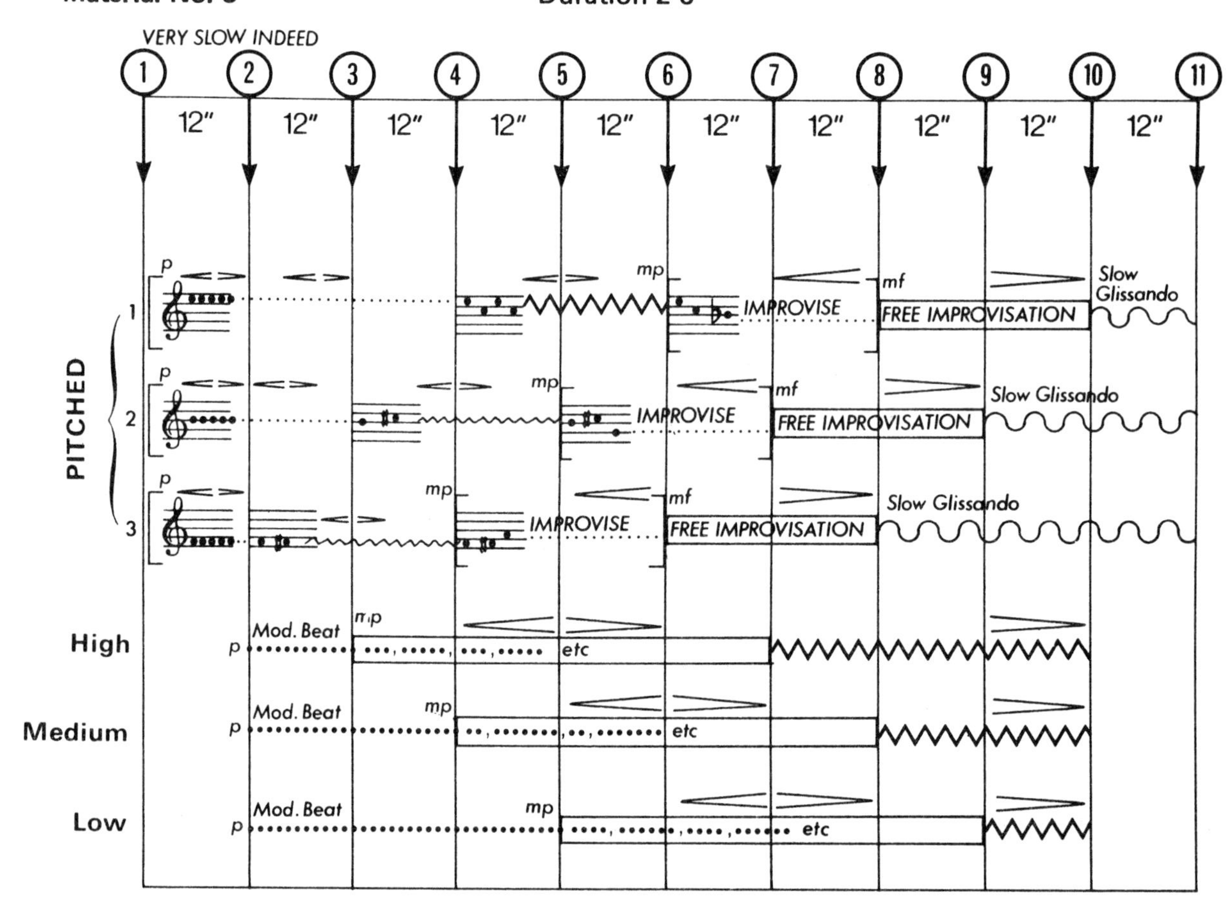

This piece makes an even more elaborate use of pitched instruments. These players should make use of the full range of their instruments in the 'free improvisation' boxes. The final glissando passages can be played quite slowly; the best effect could be for the players to make fast chromatic scales up and down their instruments. Seeing that all the material is played without a break and that transitions from one type of texture to another in accordance with the beats, players must always be careful to watch the conductor for appropriate cues.

Material No. 6

Here again pitched instruments are used with a wide ranging variety of effect. The various types of notation have been already described in this chapter and should be made clear to the players. The attack on beats, 3, 7, 11, and 15 should be very loud, especially the glissandi. The three non-pitched parts should be decided in terms of relative pitch as in the previous piece.

Material No. 6 Duration 2′0″

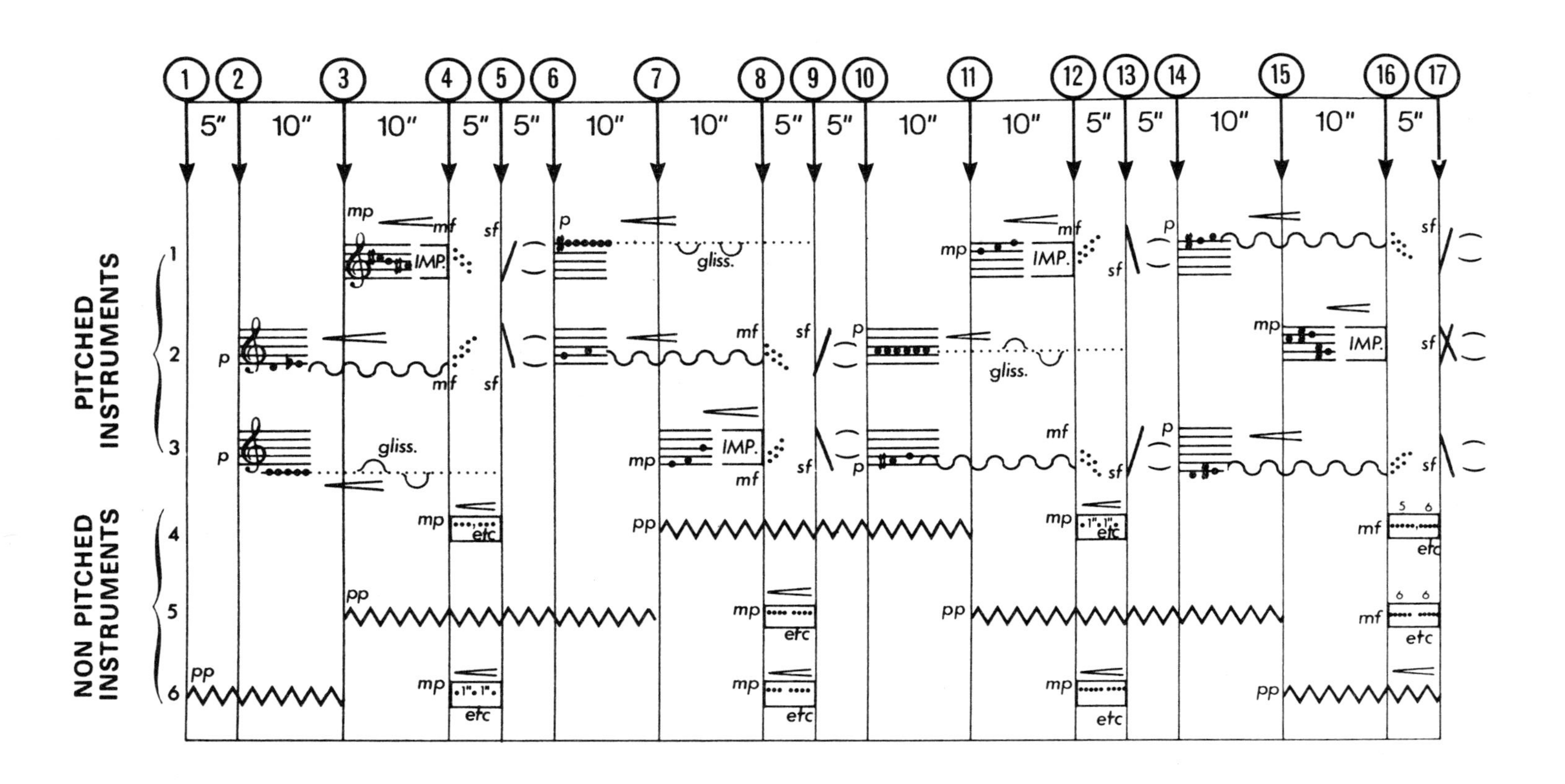

In all the pieces up to now, pulse has played a minor role. Repetitive beats and regular rhythms of one kind or another will be a prominent feature of all the remaining sixteen pieces.

The first of these (Number 7) is very simple and contains a single repeated part with three superimposed 'roll' layers. In order to maintain a consistent pulse of about • = 96, subsidiary beats can be given with the left hand, the five 'structural' beats being given with the right.

When a strictly regular beat has been mastered, the piece may be played with a marked accelerando followed by a matching ritenuto after beat three. This process can be reversed and if the piece is started with a very fast pulse, a strong rit. can be made towards beat three followed by a matching accel. so that the piece ends at the same pace at which it started.

Material No. 7 Duration 30″

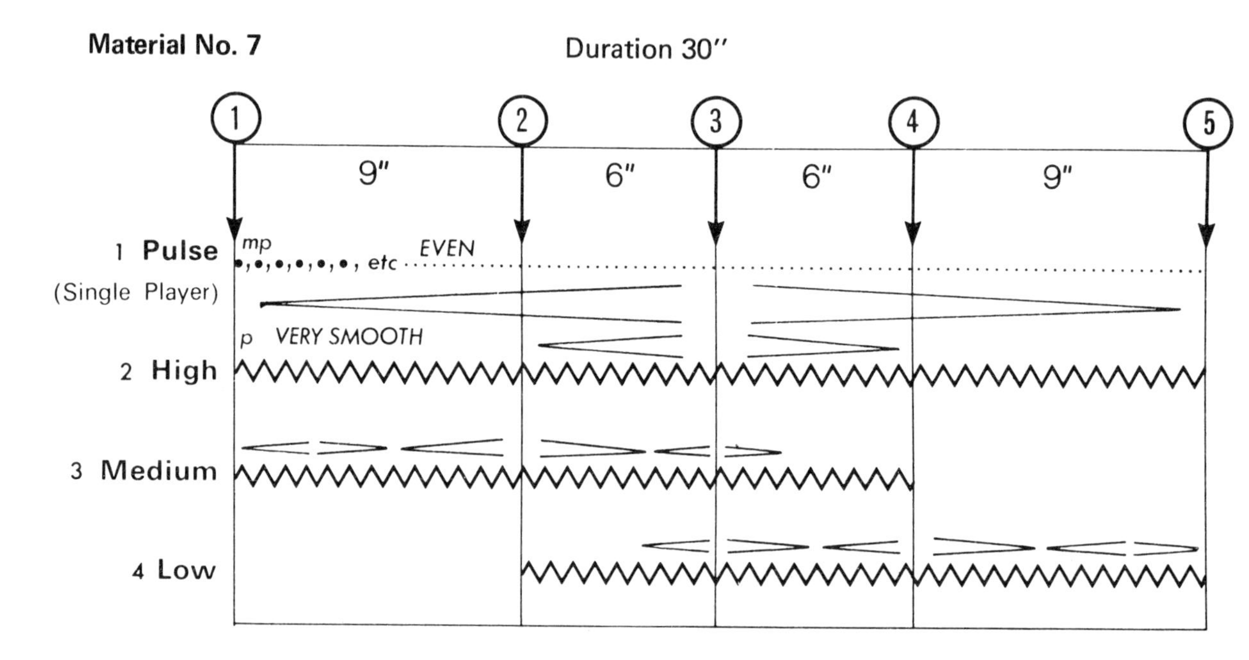

Material 8 is very similar to its predecessor and the beat part should be guided as before with the left hand, this time at about • = 72. The remaining auxiliary parts are rather more elaborate, particularly the 'counter-rhythm' of part 5. This part should for preference be played by one player only on some kind of rattle or tambourine which should deliberately attempt to disrupt the constant rhythm of the 1st part. The final rhythm of part 6 should continue to add to this desintegrating effect and may be played by several players as usual.

Material 9 relies for its effect on the synchronised damping of instruments on certain beats. Almost all the instruments should be resonant. Complete togetherness may take a little practice. The central block of rolls and rhythms between beats 7 and 8 contains several crescendos and diminuendos. These may be indicated with subsidiary gestures in the usual way (see Chapter 2 part 3).

Material No. 8 Duration 40″

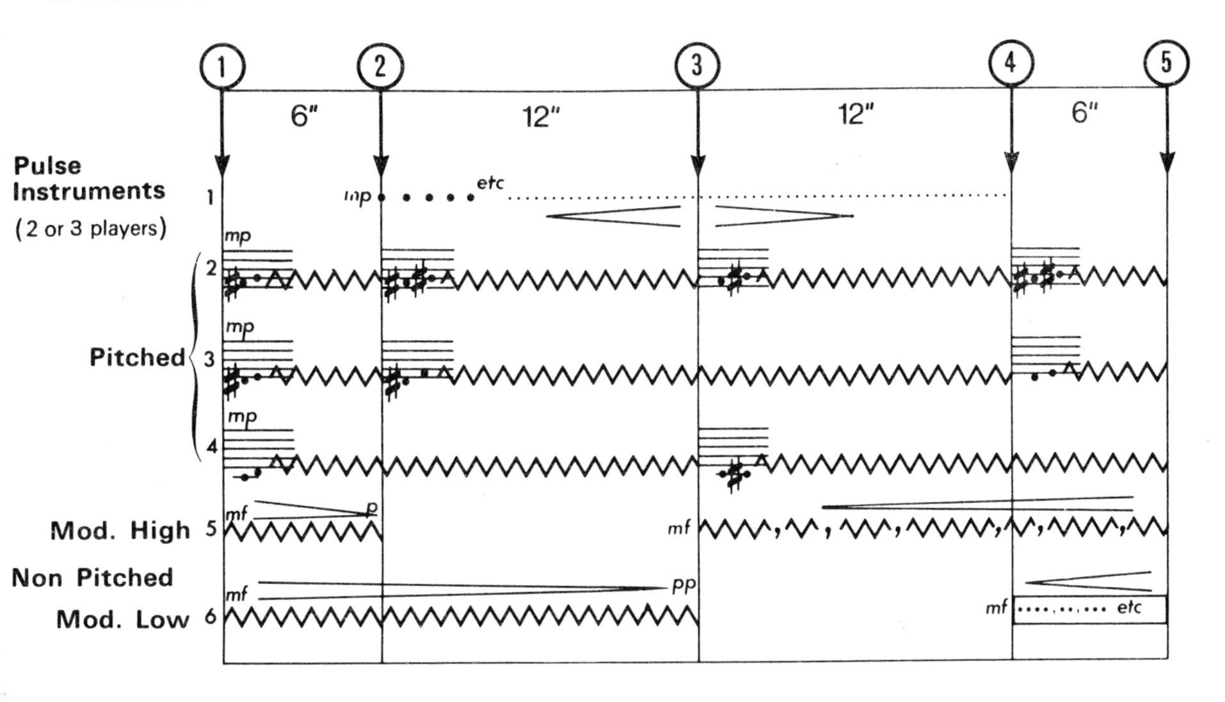

Material No. 9 Duration 1′0″

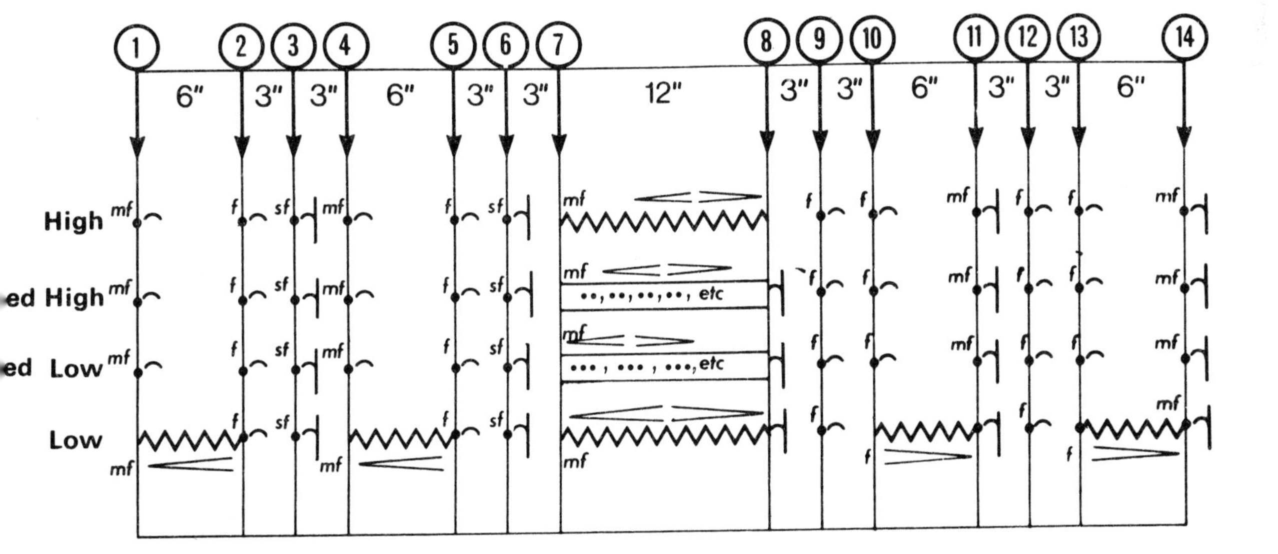

Material No. 10 Duration 1'20"

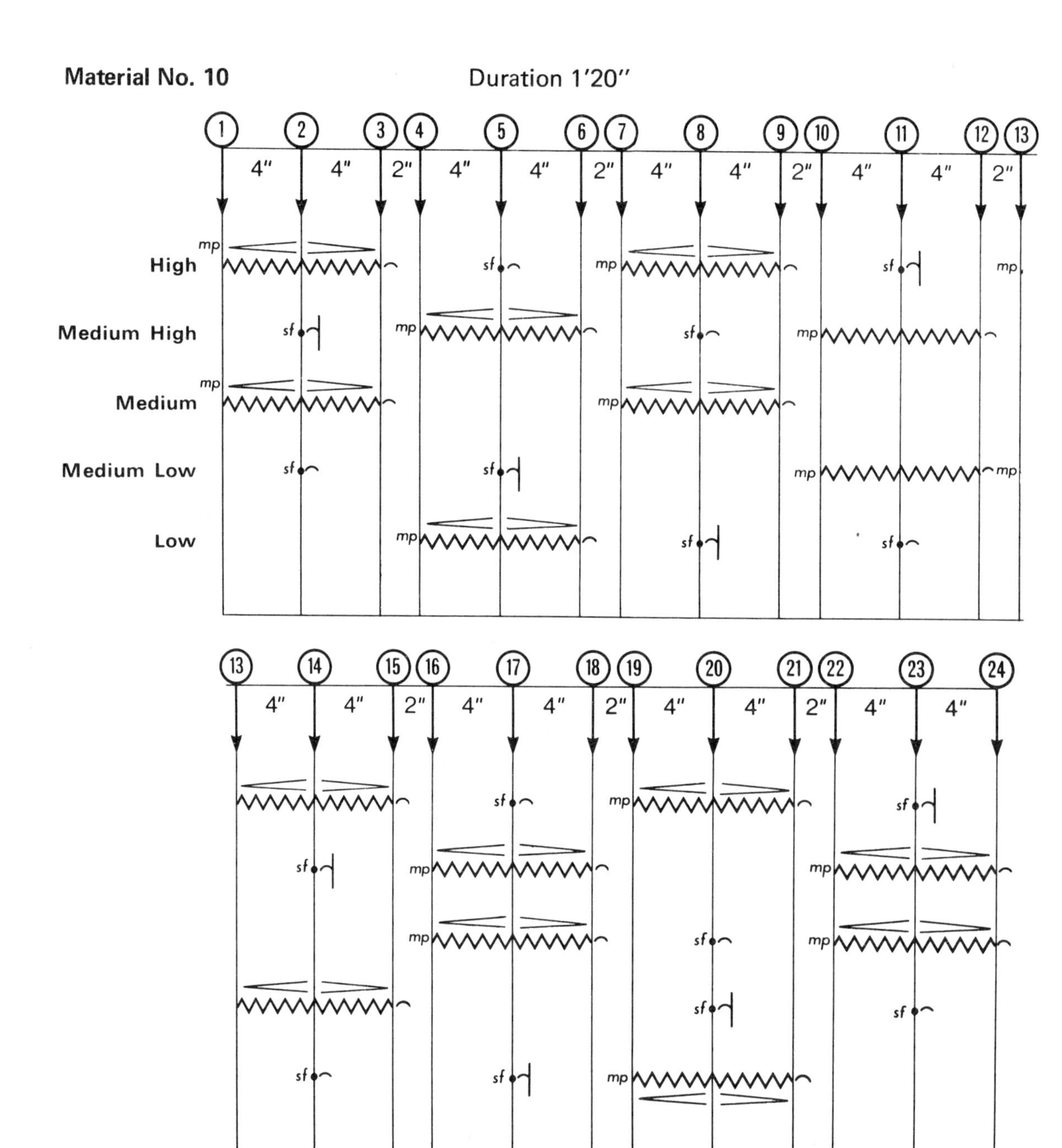

A regular wave formation provides a relentless pulse throughout this piece. In many ways it is one of the simplest materials so far. The central attack in each wave must coincide with the loudest moment of each roll pattern. Complete dynamic accuracy and ruthless regularity must be maintained as far as possible. All crescendos and diminuendos should be well marked.

Material No. 11 Duration 57"

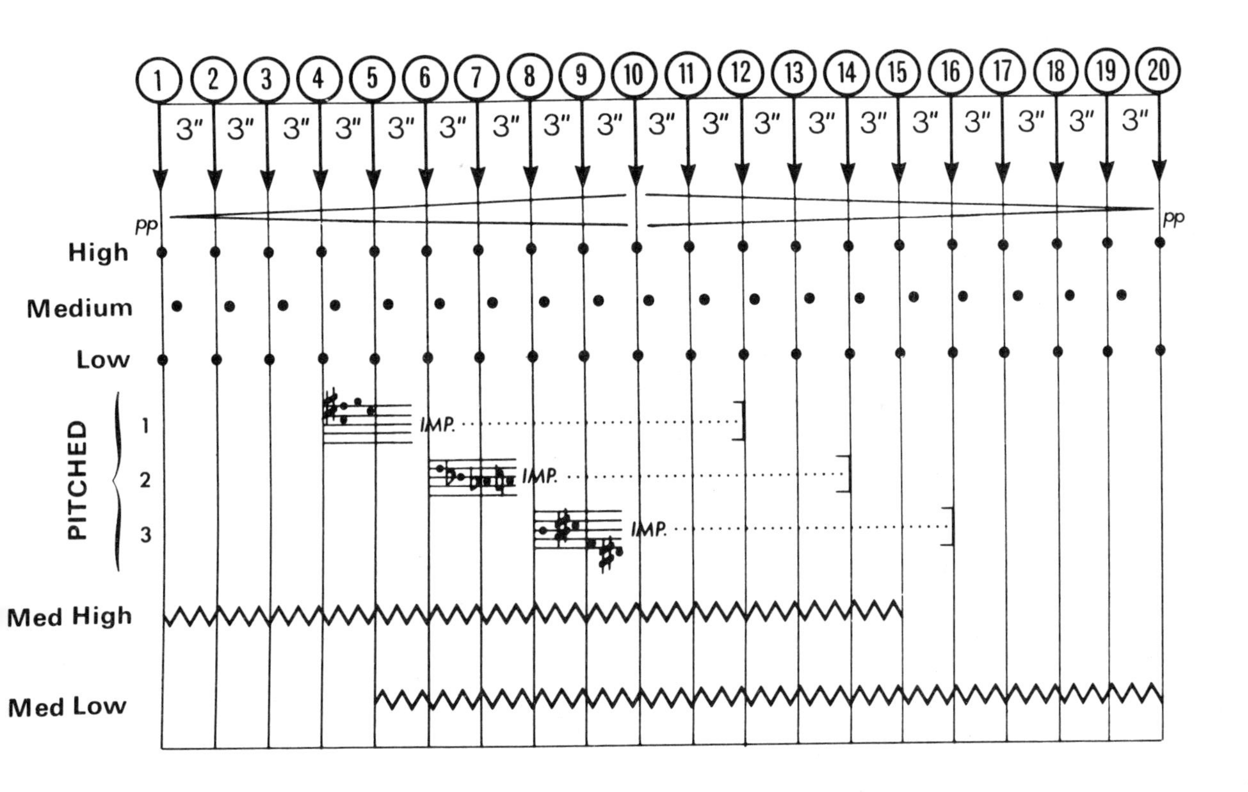

This material is in $\frac{3}{4}$ although it is not barred in a conventional way. Parts 1, 2, & 3 provide a continuous ostinato of ♩𝅗𝅥 and the conductor's left hand should help to preserve regularity (part 2 provides the second 'beat' of each 'bar'). The slow crescendo and subsequent diminuendo of the ostinato is very hard to achieve with any degree of smoothness and balance. The ostinato should be practiced separately at first.

Material 12, like its predecessor, is in $\frac{3}{4}$ but this time a traditional beat of three in a bar should be adopted and the players should count downbeats. Players 1 & 3 always play on the first beat of each bar, player 4 on the second beat and player 2 on the third. This has to be consistently maintained and players 1 – 4 should be rehearsed separately at first. Note the sudden accents in 'bars' 5, 8, 11, 14 & 17 which should be made by all of the 'rhythm' players. Just as in Material 11 the gradual cresc. and subsequent dim. has to be very carefully graded. The subsidiary roll parts (5 & 6) are very simple but, with the $\frac{3}{4}$ rhythm, counting the downbeats is by no means so easy. Relative pitches are givenfor the division of the instruments.

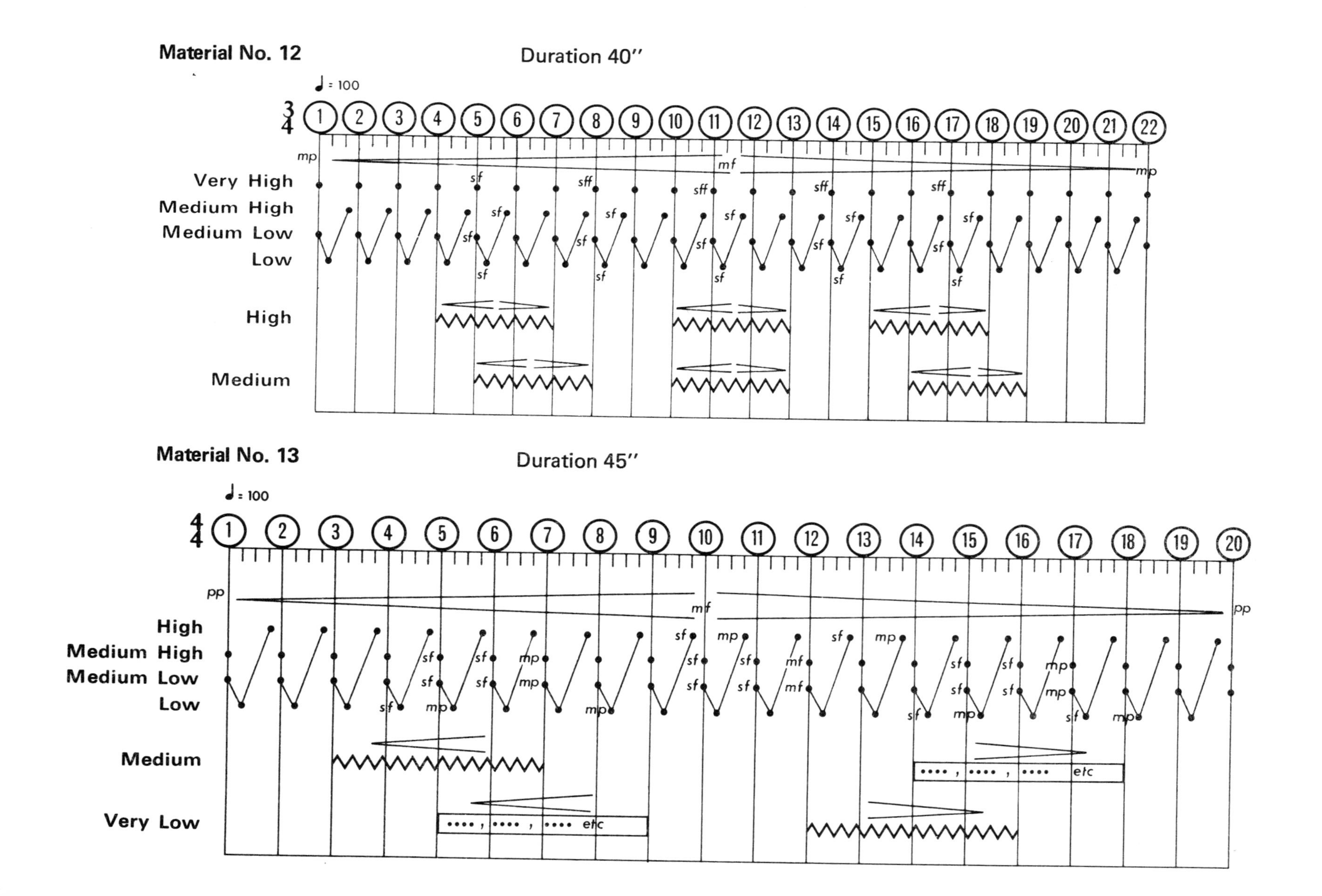

Material No. 12
Duration 40''
♩ = 100
3/4
Very High
Medium High
Medium Low
Low
High
Medium
mp
mf
mp
sf
sff
Material No. 13
Duration 45''
♩ = 100
4/4
High
Medium High
Medium Low
Low
Medium
Very Low
pp
mf
pp
sf
mp
mf
etc

Piece 13 is in $\frac{4}{4}$ which must be beaten throughout. A different instrument (or instruments) once again provides a different colour for each beat of the bar. The rhythmic pattern produced throughout should be ♩♩𝄽♩ The particular arrangement of the players is made clear in the score as in Material 12. Again players must count downbeats and this time accents occur in different parts of the bar at different times, which means that *all* the 'beat' players must count very carefully to make the accents at the right times. The rhythm boxes of parts 5 and 6 should be played 'against' the $\frac{4}{4}$ rhythm.

Material No. 14

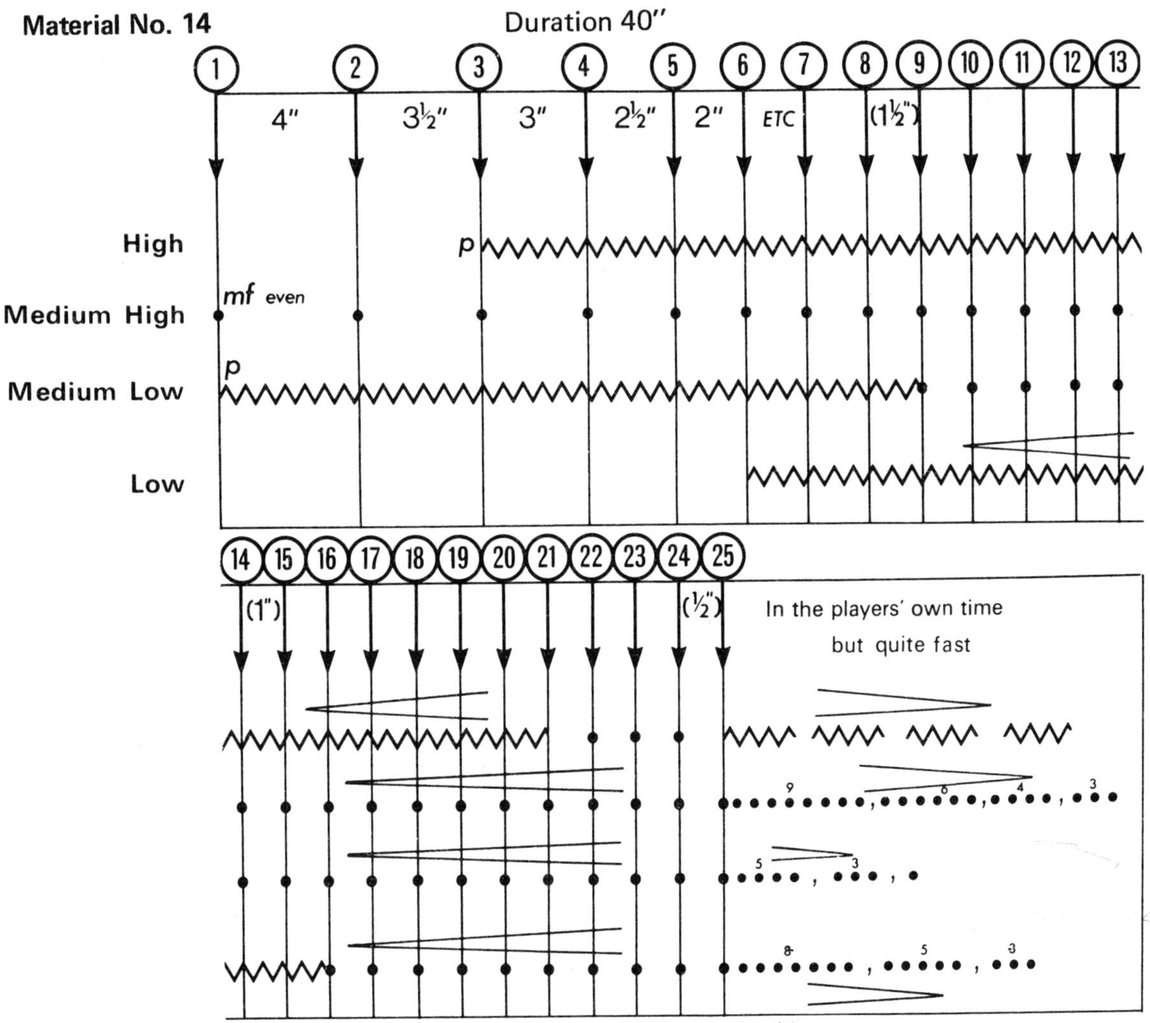

N.B. For single pulses, all instruments must be struck without resonance.
Players of part 2 should choose non-resonant instruments.

A sustained accelerando is the most prominent feature of this piece and should be accompanied by a very gradual crescendo (not marked until the end). The players should by careful not to play the materials of the final 'cadenza' too quickly – the commas between the rhythms and rolls should be reasonably long and each part should make a diminuendo so that the final outburst dies away to nothing.

Material No. 15 Duration 48″

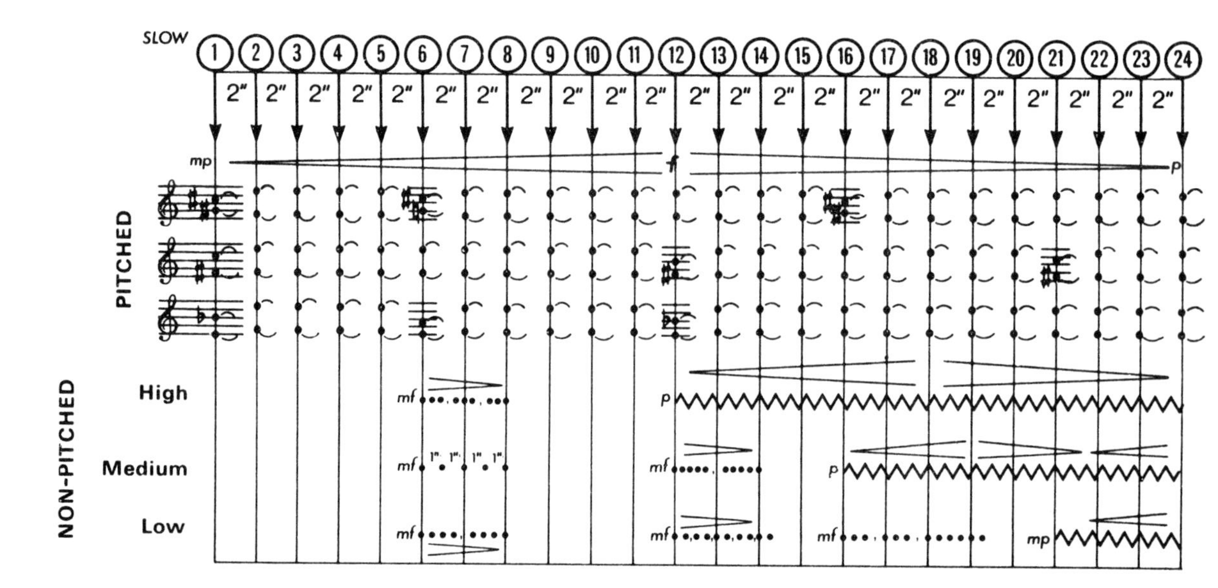

Two rather smaller 'cadenzas' are also a feature of Material 15. The continuous pulse, provided by the pitched instruments should be absolutely regular and the overall crescendo and diminuendo should be very carefully graded. The dynamic changes in the three final rolls must also be executed very smoothly and, since they move against the diminuendo of the pulse chord, should be completely differentiated from it.

Material No. 16 Duration 35″

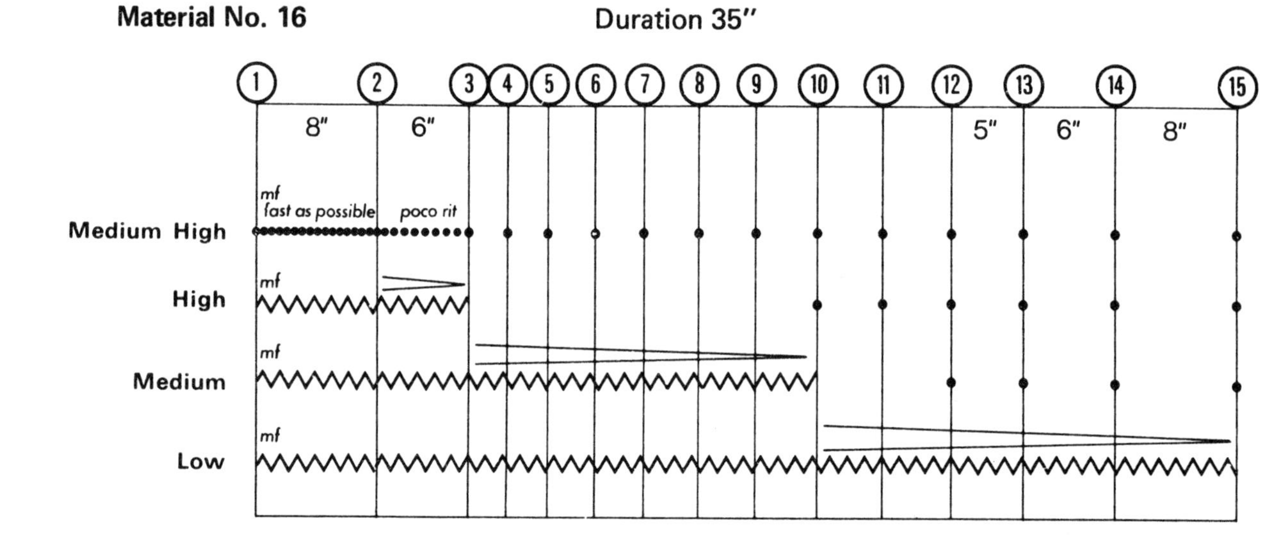

This piece is the opposite to Material 14. The opening pulse should be very fast indeed and should only begin to slow down between beats 2 & 3. The conductor must try and effect a smooth ritenuto, taking over as accurately as possible the speed of pulse just before beat 3 and then gradually lengthening the beat right up to the end of the piece. The final two or three beats should be separated very widely indeed (an approximate timing is given in the score). All the beats must be without resonance, as in piece 14, and part No. 1 should consist of non-resonant instruments.

Material No. 17 Duration 1'18"

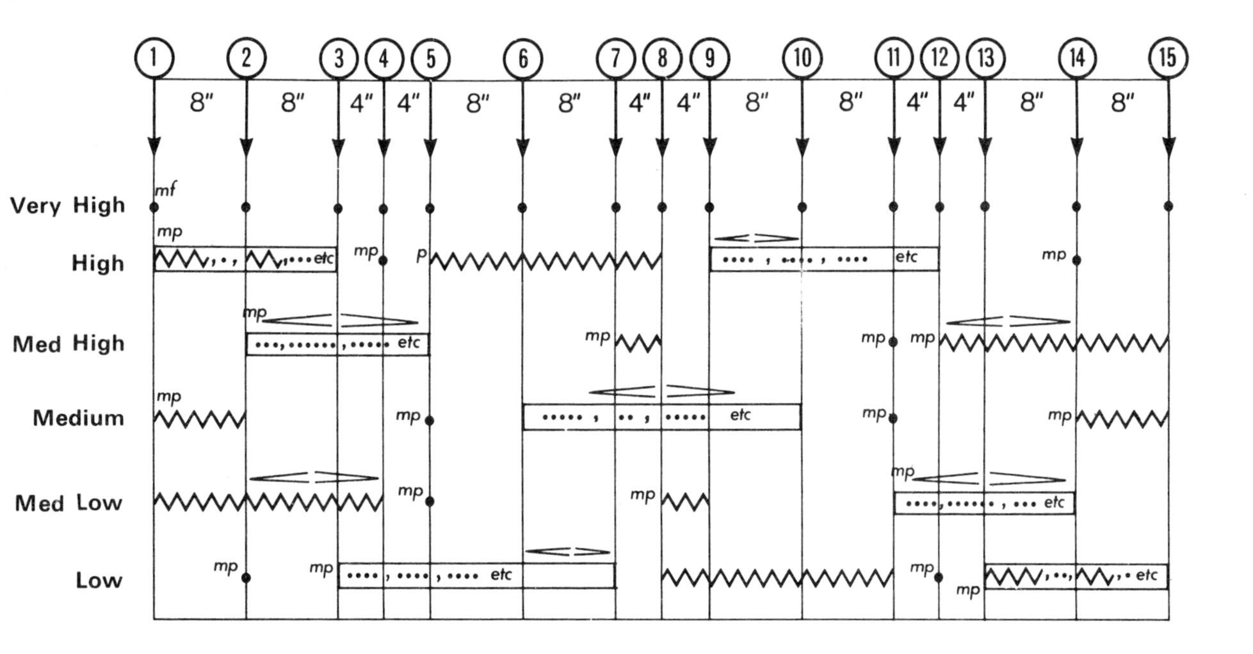

Exact synchronization is essential in this piece. The attacks which mark the changes in the blocks of rolls and rhythms should be loud and exactly together. The players with continuous material must also react very precisely with the beats so that all changes are instantaneously effected. All resonant instruments should be damped extremely quickly and this applies to the end of each rhythm box or roll as well as to the single strokes.

Material No. 18 Duration 1'40"

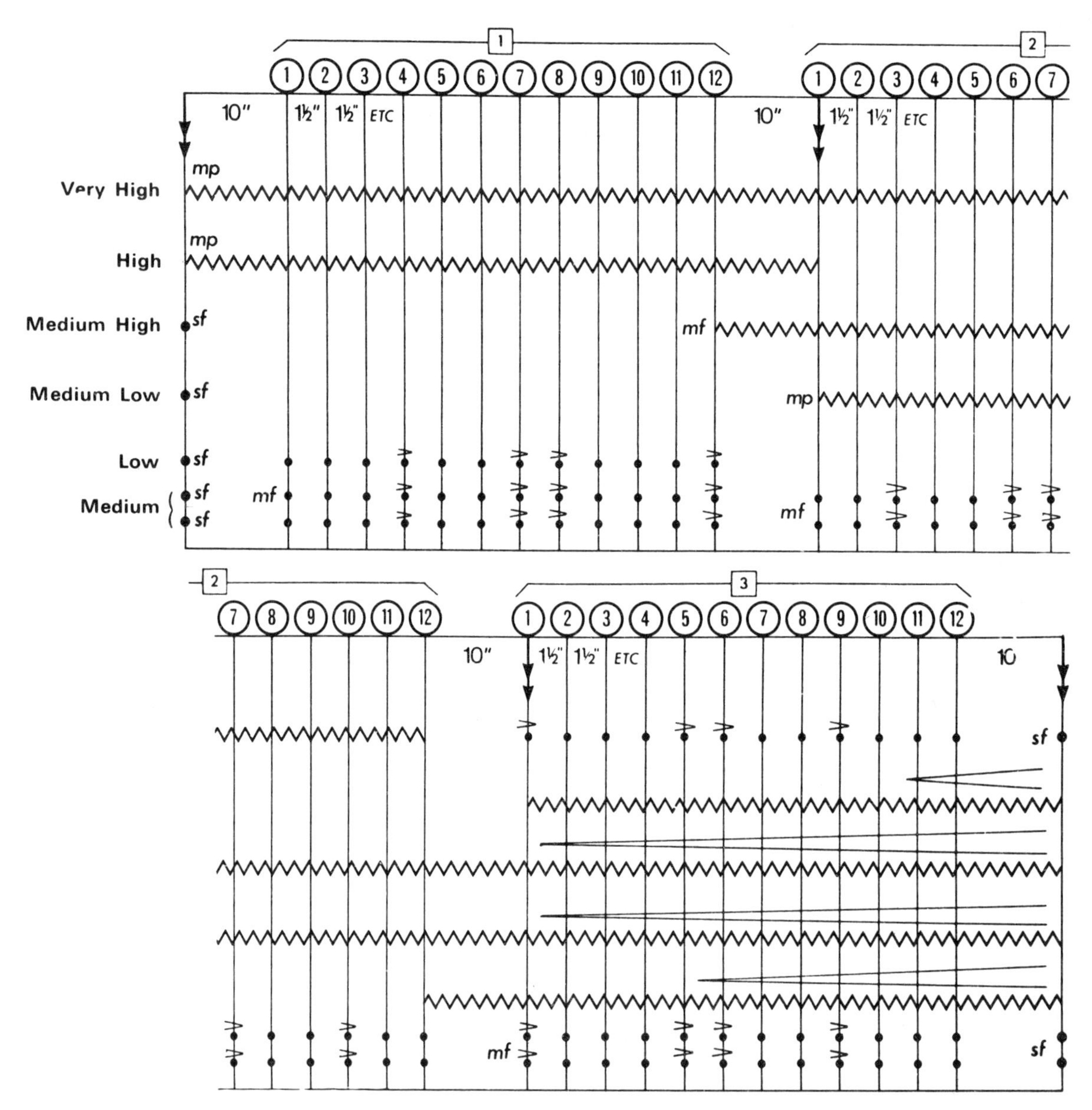

This piece has a cue to start and a cue to end as well as three sections of 12 beats in each. Since there is a 10" pause between each section, the procedure should be quite clear as far as the players are concerned and no further indications should be needed to signify the beginning of each section. The given accents should be made by all the players making beats at the given time and these must be carefully synchronized. The players must, as in the previous two or three pieces, be careful to count these fast beats so that all the accents come together.

Material No. 19

Duration 45″

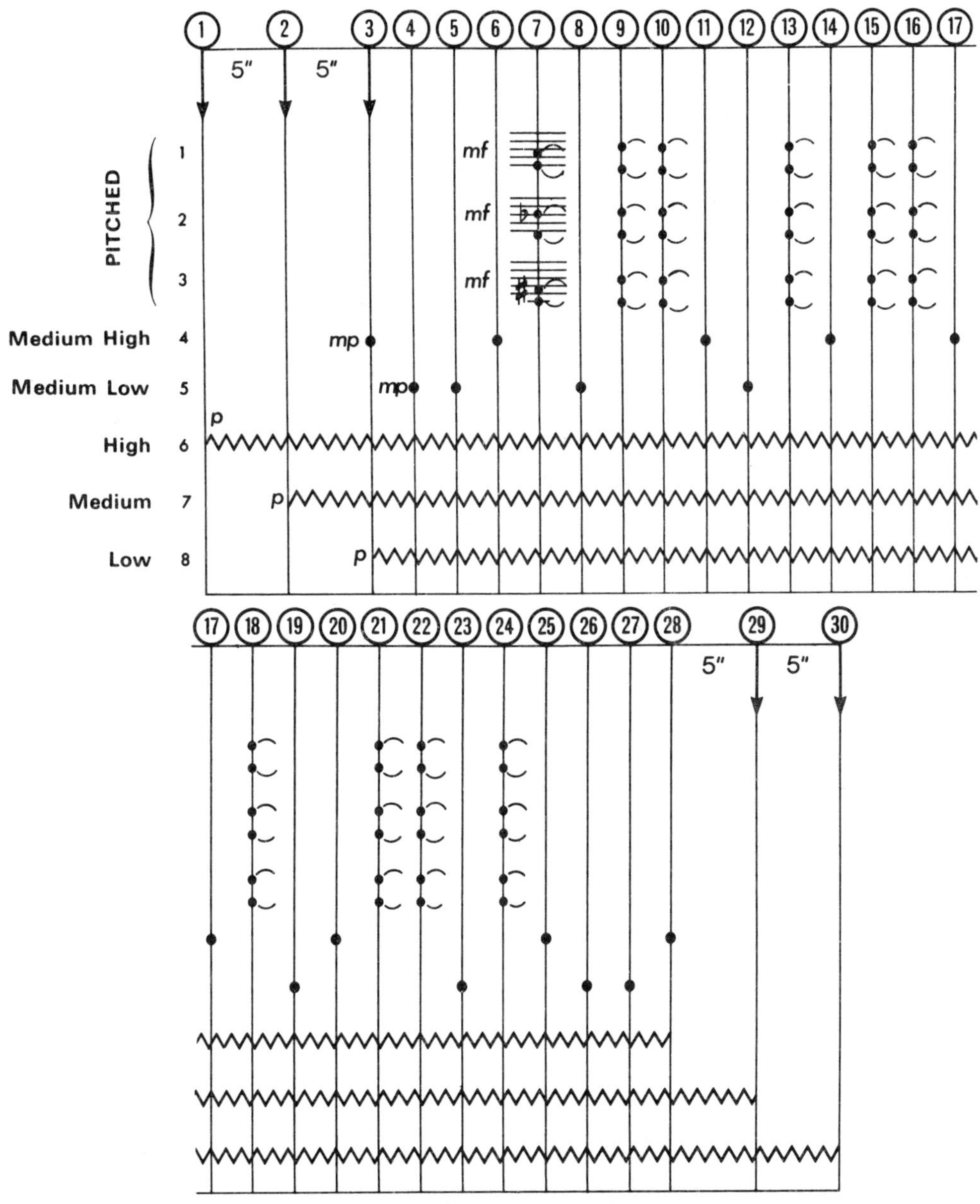

N.B. Pitched players use the same notes throughout

Piece Number 19 contains a central section which permutates a set of three different colours. The permutation is quite intricate and might require a good deal of practice. Just as with the previous piece all the beats (between 3 and 28) should be completely regular and the dynamics should be absolutely even.

Material No. 20

Duration 1′4″

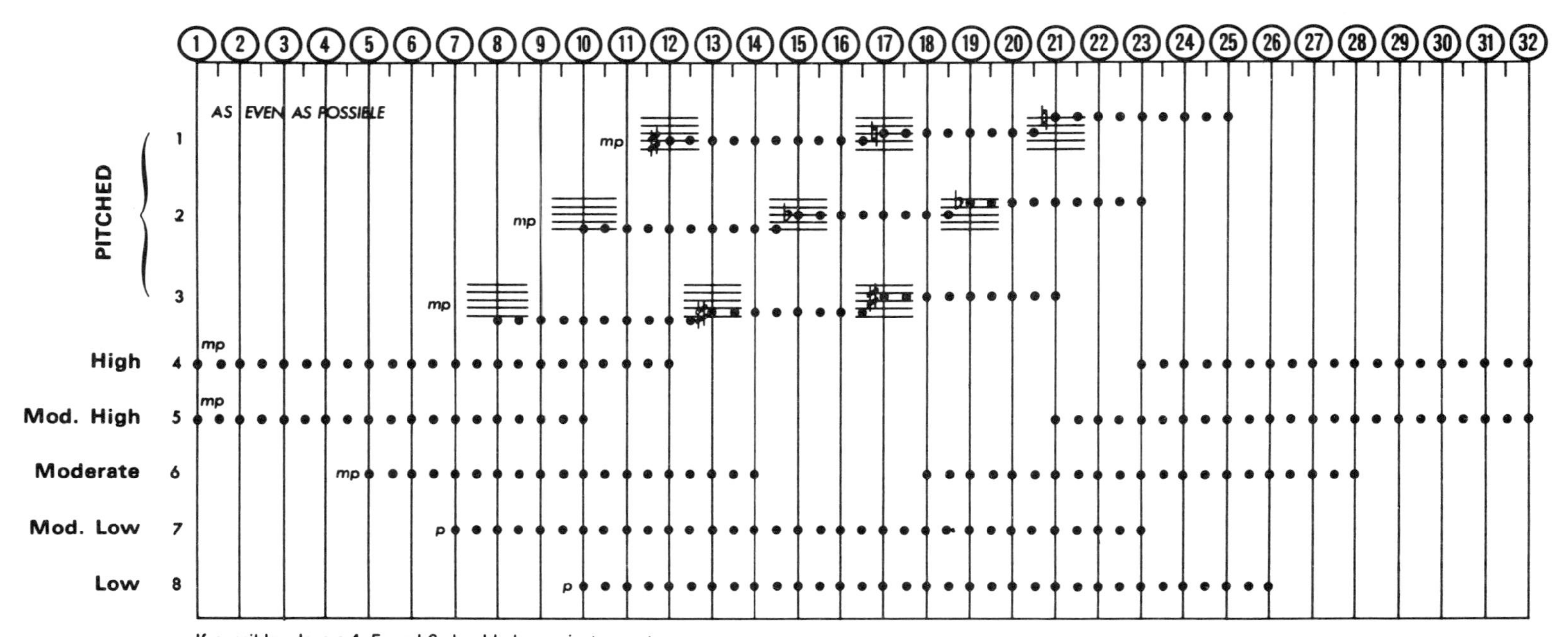

If possible, players 4, 5, and 6 should change instruments half way through, say from wood to metal instruments.

Material No. 20

This final piece should be beaten in $\frac{2}{4}$ throughout and is, in many ways, the most difficult as far as counting is concerned. The players must count each downbeat of which there are 32, The players of pitched instruments (players 1 – 3) have to change pitch at specific points and this may prove difficult in this context.

Like many of its predecessors, Material 20 is designed to sound effective when speeded up on tape. It is nevertheless a particularly exacting exercise in coordination when used as a classroom piece. As with most of the difficult pieces in this series, a slow tempo should be adopted at first for the purposes of practice.

When performing the Materials in the classroom, there are several ways in which the players could follow their parts. It should be quite easy to write some of the simpler pieces up on the blackboard in toto. The best way is to have each Material in question copied photostatically, so that each pupil taking part has an individual copy. One final alternative is to write out a copy for each separate player. This is a procedure which I myself have adopted on a number of occasions and does not require too much trouble. Here is an example of how an individual part may be copied, part 4 of Material 6:

Player 4

Beats 4 – 5 *mp* [• • • , • • • etc]

Beats 7 – 11 *pp* ⋀⋀⋀⋀⋀⋀⋀⋀⋀⋀

Beats 12 – 13 *mp* [• 1" • 1" • 1" etc]

Beats 16 – 17 *mf* [5: • • • • • , 6: • • • • • • etc]

Any periods during which the part is silent, e.g. between Beats 1 and 3, need not be written out. Durations need not be given either: they are only necessary for the conductor.

CHAPTER 4

SIMPLE CREATIVE WORK

Many of the twenty materials of the previous chapter are very simple in construction and the general principles by which each is composed are sufficiently clear for the pupils to construct pieces of their own along similar lines. Initially the pupils need not construct their music in such a geometric way, but in a freer manner. For the purposes of classroom composition, a duplicated grid could be given to each pupil. George Self has advocated something along the following lines:

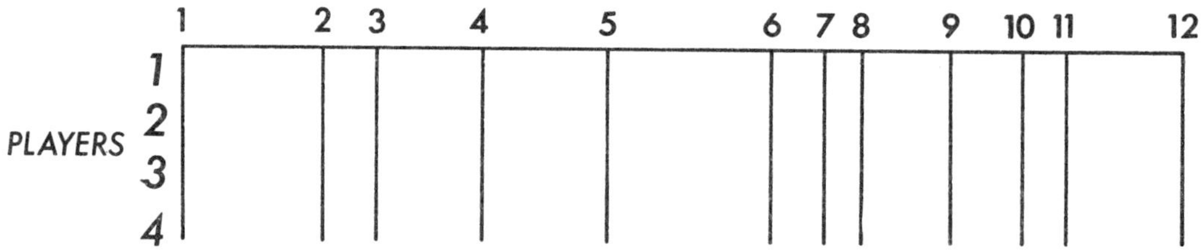

	1	2	3	4	5	6	7	8	9	10	11	12
PLAYERS 1												
2												
3												
4												

The pupils can then fill in the given parts (1, 2, 3, & 4 in this case) with points, rolls, and rhythms, and should be free to choose whatever instruments they wish. The younger the pupils the more restricted their material should be when they begin; the complete mixing of all possible elements should be the last of a number of attempts (i.e. the first step might be a use of point sounds only, then of rolls, than a mixture of both, and so on). Finally the pupils themselves can be asked to determine the positioning and number of beats and decide upon the overall duration of the piece.

There are several pitfalls which will soon become obvious to the teacher. The commonest of these is for the pupil to create too great an overall density, by giving too much for each of the players or groups of players. A gradual build-up and its opposite, the thinning out of layers, should be encouraged. A second common mistake is the opposite of the first, the creating of a very sparse texture with one sound tending to follow the other without *any* overlapping of parts. Some sort of balance should always be maintained between density and simplicity and this sort of problem should be discussed at length with the pupils. A certain pupil might be quite adamant in keeping to one of these two extremes, feeling that his selection of materials was particularly suited to this kind of treatment. The invention of striking mixtures of colours should always be encouraged. The pupils should be asked always to specify dynamics and to vary constantly the ways in which the instruments should be played. The canonic and other geometrical patterns found in the pieces of this book can be imitated and used as a basis for shaping more disciplined materials. Finally, it is extremely important that as many of these pieces as possible should be tried out in the classroom, so that the various symbols can be translated into sound.

Here is an example of the sort of piece that might be produced by a pupil. In this case an ordinary piece of lined paper was given to each pupil; this was then laid horizontally and divided into two, making two sections of equal length. Each line on the paper was to represent

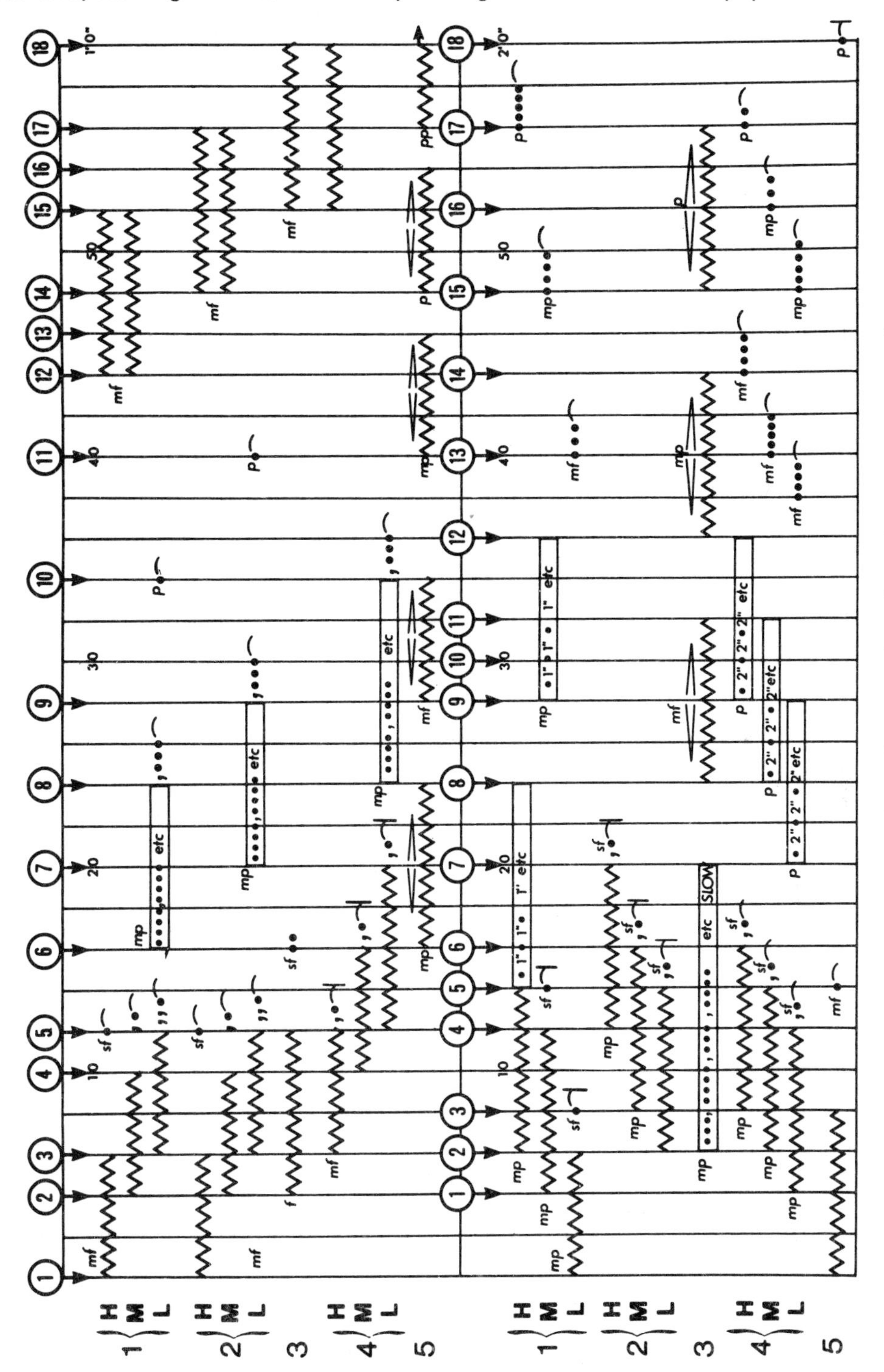

Piece by Keith Ruskin, pupil at Shoreditch School, aged 13.

about two seconds. The pupils were told to select five instruments or groups of instruments, and in this case the following were chosen:
part 1 – three glockenspiels (high, medium, and low)
part 2 – three groups of chime bars (high, medium, and low)
part 3 – marraccas
part 4 – three cymbals (high, medium, and low)
part 5 – low drum(s)
This piece has in fact eleven real parts although I had only expected five. The division of parts does however show a welcome degree of initiative and makes the piece much more complex in effect. The number of cues and the placing of the cues were decided entirely by the pupil himself and the fact that there are an equal number of cues for each of the two sections of this piece is fortuitous.

2

Strong visual elements can be applied to other aspects of composition, not just to the overall structure of overlapping parts as above. For instant, geometry if used in connection with rhythmic pattern-making, can be applied at a much earlier stage. A grid such as the following could be given to each pupil (8 x 8 is suggested because it allows for easy calculation);

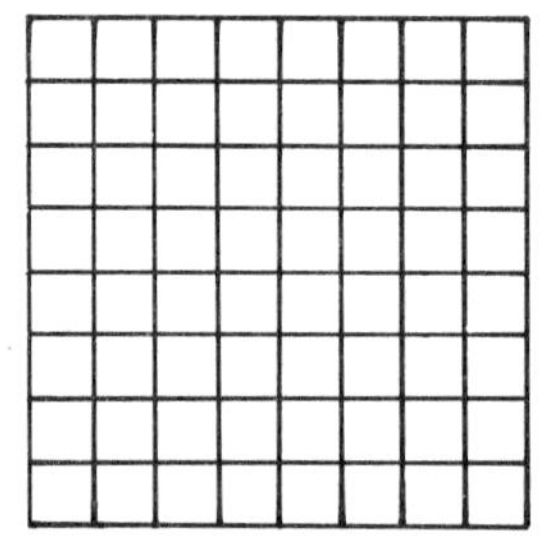

Each pupil should then be asked to fill in the spaces with either single beats (•) or short rolls (/\/\) leaving occasional blank spaces to create pauses. The result should always be conceived in terms of the resulting rhythm (reading the pattern from left to right, moving down one line at a time) while at the same time a pattern which is visually effective should be encouraged. A strong visual pattern in fact often implies a complex rhythmic pattern. Here is an example by an eleven-year-old:

Ex. 1

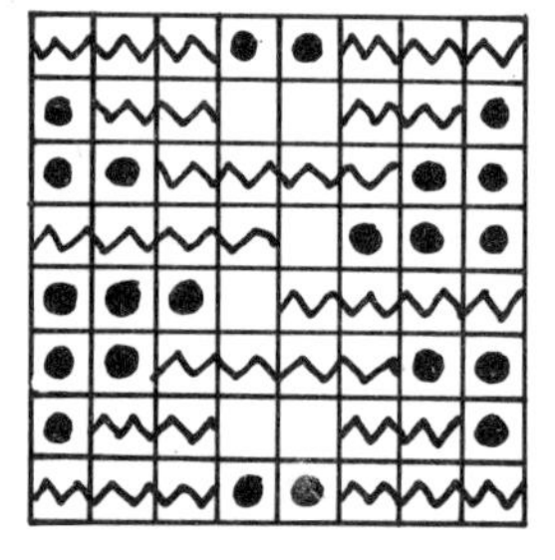

The fact that the second half of this rhythmic square is the exact retrograde of the first was not due to any form of indoctrination in serial methods. Making mirror images is just a simple way of achieving symmetry.

A further complication can now be suggested by allowing two beats to come together in some of the squares. This will imply a doubling of the rhythmic pulse. Here is another example by one of my first year pupils:

Ex. 2

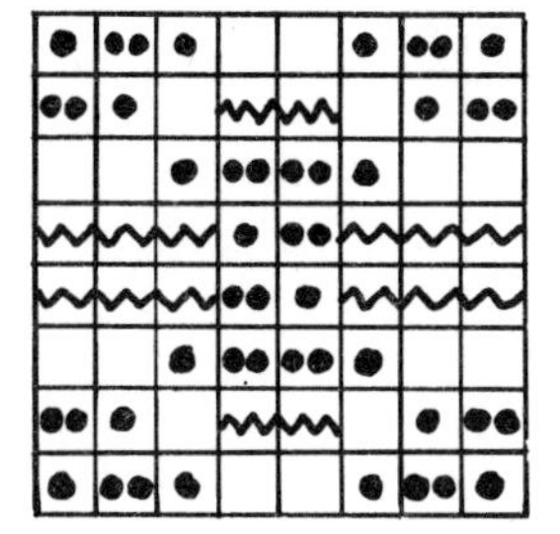

Now Groups of three or four beats can be introduced. The last line of the next example, also by a first former, could be written out in the conventional way as follows (taking one crotchet per square):

Here is the entire example in its original form:

Ex. 3

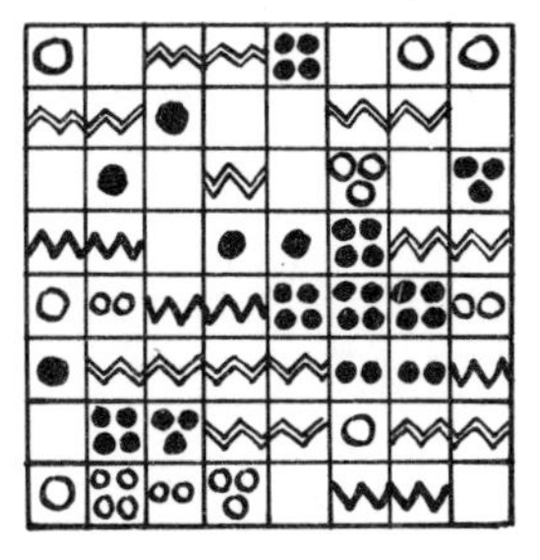

= blue crayon in the original.

= red crayon in the original.

Finally the whole process can be made more complicated still by substituting numbers for groups of beats. Rolls can be made more interesting by writing them obliquely so that implies a speeding up and a slowing down of the given roll.

Both of these next examples are by second formers:

Ex. 4 and Ex. 5

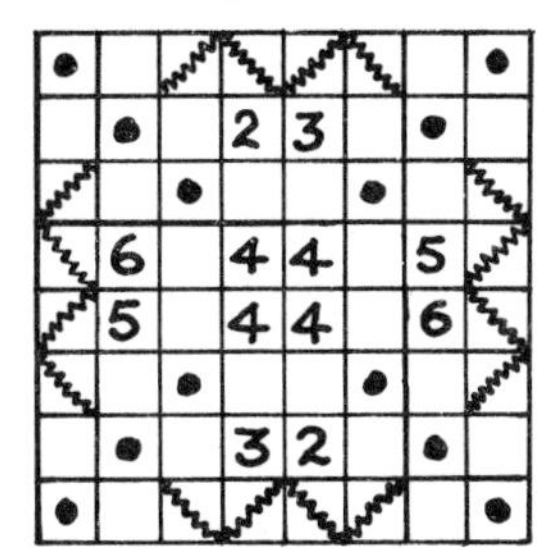

3	4	5	6	6	5	4	3
	3	5			5	3	
	4	6			6	4	
8							8
8							8
	4	6			6	4	
	3	5			5	3	
3	4	5	6	6	5	4	3

In order to clarify the implications of the number system, here is a realization of line 1 of Ex. 5 in terms of traditional notation:

It will be agreed I am sure that the original is considerably easier to follow, saves a great deal of time in writing out and provides a much clearer overall impression as a structure.

The various symbols suggested need not be taken as exclusive of any which the pupils might like to invent. For example one of my pupils used dots of varying sizes and rolls of varying amplitude to suggest loud and soft. I also noticed that several pupils used coloured pencils and it was agreed that these different colours could represent different instruments, but the pupil may be able however to extract a sufficient number of contrasted sounds from one instrument.

All these rhythmic patterns should be tried out on an instrument or instruments as soon as practicable. Writing down and playing should become almost a continuous process.

Experiments can be made by realizing the 'magic' square in various different ways. One pattern could be read like Chinese writing, moving downwards in rows starting from the extreme right of the pattern. One rhythm might be read backwards starting from the bottom of the square or perhaps in chain formation as follows:

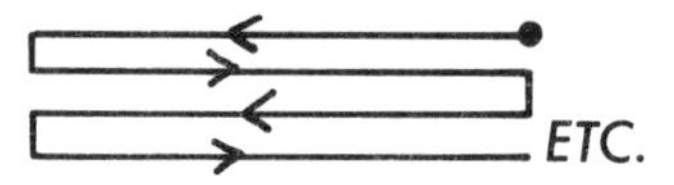

A series of diminishing squares might provide yet another intriguing route to follow:

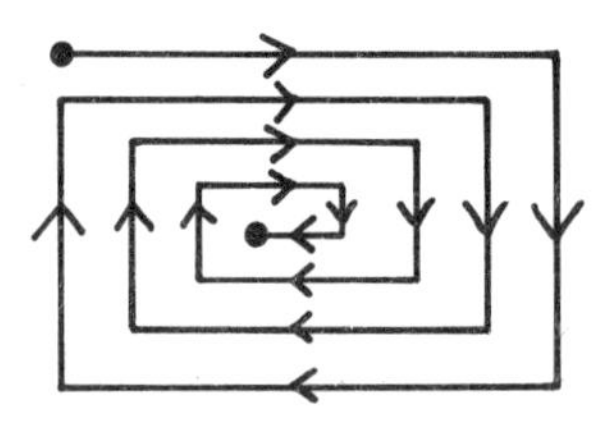

Finally one can combine two or more players to provide a whole series of rhythmic canons and other forms of counterpoint. Keeping together may be quite a problem for the players when this is first attempted, and if previously the pupils have tended to interpret their rhythms rather freely, a greater consistency of rhythmic pulse will be necessary when parts are combined in this manner.

These invented rhythmic squares could provide an alternative source of material for the 'rhythm' sign improvisations described in Chapter 2. The conductor could in this case either stop a given rhythm in mid-field or allow it to be played through in its entirety, in which case the player concerned stops of his own accord when his rhythmic square reaches its end. With sufficient cunning the conductor could build up layers of rhythmic patterns which are related to the natural divisions of the squares themselves. In other words he could take the first rhythm he initiates as his base and cue in subsequent rhythms in accordance with the number of lines reached by this player.

Like the majority of ideas in this book, the above method (Magic Square rhythms) may be modified at will by either the teacher in question or by suggestions from the pupils themselves. The specified dimensions (8 x 8) or even the square shape do not have to be rigidly adhered to. Both the teacher and the pupils should always feel at liberty to experiment with freer shapes (extremely irregular shapes if desired) and correspondingly freer patterns.

3

A less restricted kind of pattern-making can be readily applied when dealing with pitched instruments. As a start one can simply encourage the children to create patterns or figurations which might suggest musical sounds to them. Most of these, I have found, can be realized on a single instrument, preferably with a wide pitch range, such as piano, organ, xylophone, or glockenspiel etc. Some shapes may well be suited to one instrument in particular. Some may be impossible on any instrument. These are the sort of things which can be discovered by experimenting and discussing the result. To illustrate this here is a group of interesting shapes drawn by a second-year pupil as a first attempt to create pitched sounds. I have given each shape a letter for reference.

Ex. 6

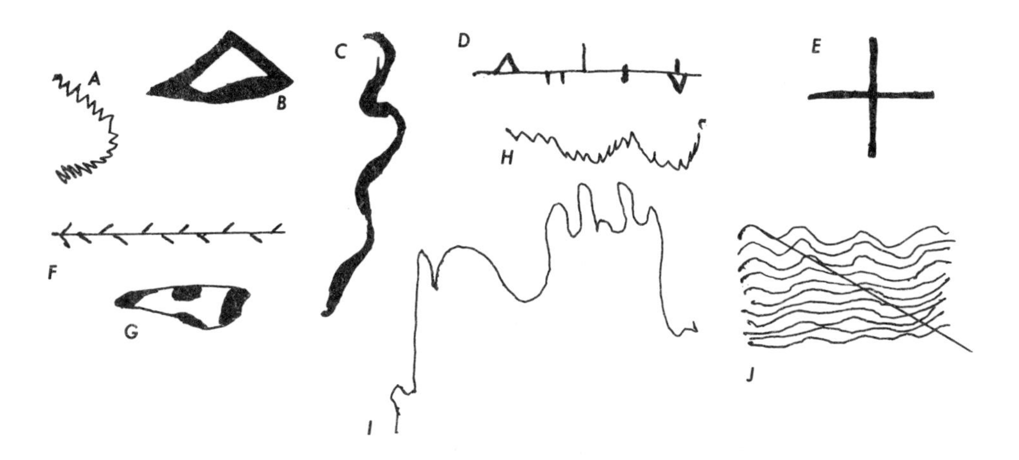

The most suitable instruments for realizing these shapes should be discussed and if the instrument or instruments are available, the teacher might first of all demonstrate a sound — or rather realize a pattern — and then allow the pupil or pupils to try for themselves.

Shape *a* might be played on a xylophone as two converging oscillating glissandos i.e. one should trill rapidly with each stick whilst moving inwards.

A shape like *b* is best on an organ and can be played as a cluster of notes slowly accumulating, then gradually splitting in two with the right hand moving upwards, and then with both hands reconverging in contrary motion. Clusters can move upwards or downwards by slowly adding pitches whilst releasing others. Shape *c* of this example is impossible unless it is turned on its side: a musical shape cannot turn back on itself because time only progresses forwards. On its side this figure can be realized as a melodic shape, a slow glissando or a moving cluster. Almost any pitched instrument could be used.

d, e and *f* can all be realized as cluster shapes, each with a sustained pivotal line. In *d* the clusters are vertical and short; in *f* each 'branch' is a short glissando. More than one instrument could be used for *d* and *f* — two or three glockenspiels for example (one for the shapes above the line, one to sustain a central tremolo, one playing the shapes below the line).

g is an interesting shape which would need careful realization, perhaps on organ or piano. Like *b* it is best played slowly to give it plenty of room. Note how single lines (single notes possibly) emerge from the upper and lower extremes of the opening cluster and from the two central clusters.

h is a simplified version of *a*.

i may be seen as a wild, far ranging melody line or as a dramatic glissando pattern. Like *c* it does have a tendency to bend back upon itself.

Finally the dense texture of *j* seems to demand several instruments. Why not try voices? A dense oscillating chord could be produced (11 parts are indicated) with a single voice glissandoing dramatically from the top to the bottom of the register.

These are just some of the possibilities inherent in one group of shapes, as well as some of the impossibilities. When the latter in particular have been discussed and eliminated, and when each shape has been tried out instrumentally, another series of shapes can be devised which might benefit from all this. If he wishes the pupil might now devise some means of suggesting relative pitch. After all most of the shapes of Ex. 6 could have been realized at any pitch level and within any limits, wide or narrow. Now, the range of whatever instrument is used could be divided into high, medium and low. The groups of horizontal lines in certain shapes of this example can be realized as chords rather than clusters.

Ex. 7

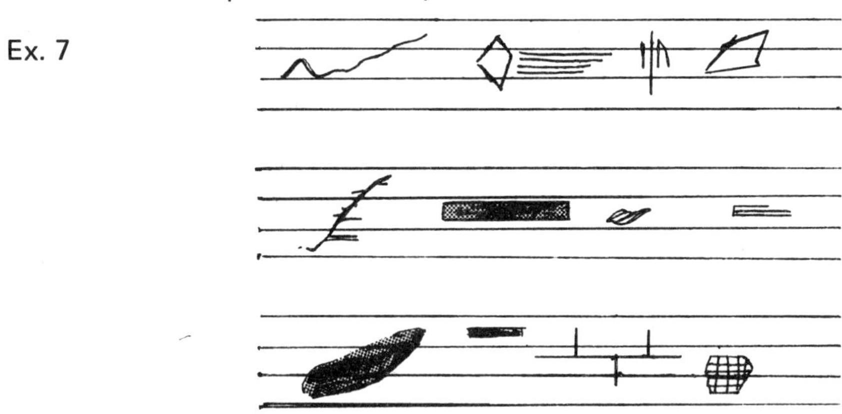

One must be careful in realizing some of the shapes above; those which have been shaded in one can normally regard as clusters; those which have not one has to interpret as non-resonant glissandi, single note patterns etc. Some new meaning has to be found also for the criss-cross shading of the final shape or the oblique shading of shape 7. Shape 5 could sound very interesting if, on the piano for instance, the horizontal lines are interpreted as a held chord emerging from out of a loud glissando, i.e. one or two appropriate notes are held as the glissando rushes upwards unsustained.

The next example is of an attempt to produce something more continuous. The vertical divisions are intended to indicate equal lengths of time. The biggest drawback, although this in itself makes an interesting point, is that pupils tend to distribute their material rather evenly within the rectangular divisions as in this example. This means that in creating such shapes, there is a natural tendency to relate them to any underlying pattern which may have been drawn, in this case the two sets of guide lines, however continuous the movement is intended to be. This particular pupil has on the other hand been quite careful in creating a material which can be played by two hands at a keyboard.

Ex. 8

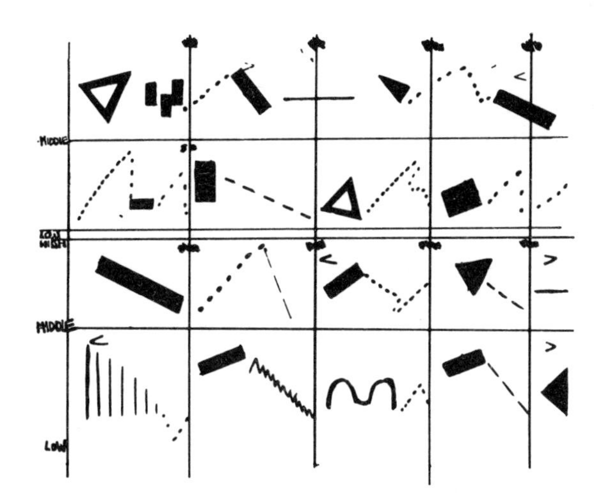

In order to solve the problem of even distribution, a grid such as the following may be suggested, with a certain number of rectangles eliminated prior to composition. The number or suggested number of these silent spaces might be agreed at the onset, but the pupil can decide their exact placing.

HIGH			*SILENT*		*SILENT*	*SILENT*
MIDDLE	*SILENT*			*SILENT*		*SILENT*
LOW	*SILENT*	*SILENT*			*SILENT*	

Another solution is to have at least one portion or layer which can act more as a background part. In the next example, the slow oscillating bass provides a neutral contrast to the dramatic, detailed nature of the upper part. Such a piece could be realized as a duet (piano and organ perhaps) with the lower part in the organ.

Ex. 9

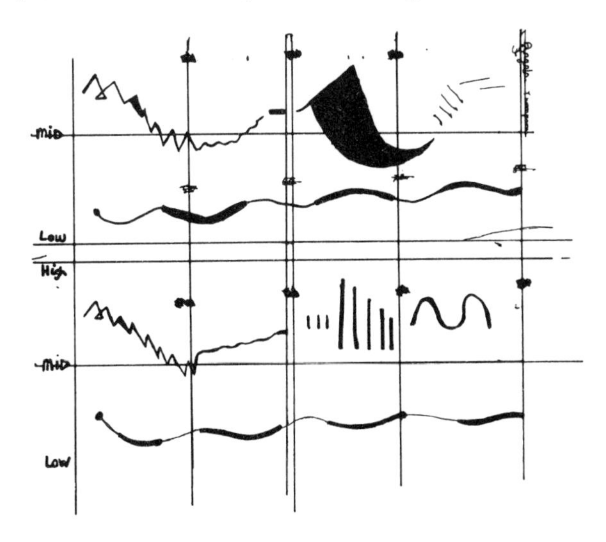

Once this stage has been reached, the pupils can, if they wish, increase the amount of detail and specification in their pieces by adding such things as dynamics, precise timings, different ways of playing a given instrument, pedalling, etc. They can also increase the number of different parts, bearing in mind that simple rhythmic parts should be given to non-pitched instruments, simple pitched parts for narrow-range instruments and more complicated parts for instruments with wide ranges. Ultimately, of course, they might produce more sophisticated examples of the kind of piece suggested in part 1 of this chapter. The geometrical or pictorial aspects of their work can now apply to individual parts and to the shape of the overall structure as it appears on the written page. There is however enormous value in realizing simple shapes and this stage should not be passed over too quickly.

It is true the realization of such shapes can often be very subjective. I myself tend to realize such shapes in a straight-forward, literal manner. For example ◀ would to me imply the gradual building up of a dense note cluster, starting with one note and adding chromatically one or two notes at a time. The following I would realize as an arpeggio: . . • • ˙ (See Ex. 8). If the notes are equidistant, as here, I would try to keep them the same interval apart e.g. a major third. Zig-zags or wave forms I would realize as trills or ostinatos, and so on.

The pupils are very likely to have their own ideas; I feel that they should try and justify them, to put over logical reasons for any unusual methods they may have. In this way an interesting, if partly subjective, relationship can be established between visual shapes and sounds, which can in itself encourage new ways of looking and listening. Music in this way can benefit from the art lesson and vice versa. The opposite process to the above, which might well be of great value to the potential painter, would be to encourage the realization of certain sounds or groups of sounds in terms of shapes and colours – slow gong strokes, glockenspiel trills, sustained chords on the organ and so on. The result would have to be judged in terms of its shape, the balance of its parts, its colour scheme etc. Only in one way would it relate to music, and this would depend on how well it could conjure up the original sound, how well it suggests the music texture, the musical shape.

4

All the ideas suggested so far in this chapter have involved some form of notation; they have been concerned with the relationship between visual images and sound. They have also involved very little collective work. Each pupil has been asked to create individually. A piece such as the first of the examples in this chapter does need quite a large number of players, and any pieces which are written in this manner should be tried out with the necessary number of parts, but this kind of participation must be adapted to the precise wishes of a single pupil. The final approach which I will illustrate in this chapter will be the making of group compositions. This form of activity can promote lively discussion, can combine the varied talents of an entire class and need not involve any written work unless specifically desired.

The making of final decisions throughout this process should be reached by majority agreement, although the teacher should always be ready to give as much advice as is felt to be necessary and also be at hand to prevent too much wrangling over minor issues.

The procedure which I invariably adopt when embarking on a class composition is as follows:

Stage 1

I first encourage the pupils to think of a suitable central idea which can then act as the basis of the piece. It is possible to have more, but one such idea is usually enough. The central idea very often gives the piece its title automatically; it should at least suggest a title. If it is a really good central idea, it will conjure up a whole range of interesting sounds and might even suggest structural and textural ideas. This is not to say that it must be specifically related to something one can hear. A sound picture can often be successfully derived from a different image altogether. Here are a few examples of basic ideas which have appeared in some of my lessons. The first three of these suggest sounds more or less directly; the latter three have broader connotations. In the fourth and sixth the basic idea provided its own title. The other ideas are accompanied by titles which have been suggested subsequently.

1. Thunder and lightning – *Storm Music*
2. Sounds of a motor race – *Le Mans*
3. Sounds in the jungle – *Bongo*
4. *Rust*
5. Card games – *Jack of Diamonds*
6. *Gemstones*

If there are a number of ideas put forward in a given lesson, a vote should be taken to decide which to adopt. Any good ideas which are not accepted can always be kept in reserve for another occasion.

Stage 2

The pupils are now asked to think of a number of sounds which best illustrate the chosen idea. The only limitation should be the feasibility of making such a sound in the classroom. The pupils should be allowed to draw on whatever instruments there are available in the school plus any other sources of sound which they themselves might like to acquire or invent. The pupils must decide on the number of different types of sound they wish to have and if a given idea is particularly suited to a subtle and limited range of instrumental colour, they must decide when to call a halt. In other words they must decide on just the right number of instruments needed – not too many and not too few. When a selection has been made, a few preliminary suggestions as to how a given instrument might be played should now be put forward. This will particularly apply to an instrument like a suspended cymbal from which countless different noises can be extracted, each with a different associative meaning. Most of the subtleties of performance should however be left to the end as a final tidying-up process. Only those considerations which affect the overall sound significantly should be discussed.

Stage 3

The basic idea must now be carefully scrutinized to see if it can suggest some kind of structural pattern. To illustrate this process I will take the six titles given as examples in Stage 1 and make some suggestions for each.

1. *Storm Music*
This piece could begin quietly (in the distance), reach one or more highly explosive climaxes and then die away again. There might be three basic sound elements to illustrate three of the main components of a storm – i.e. thunder (drums, wobble boards etc.), lightning (cymbals, triangles, and other high pitched metal sounds) and rain (marraccas, shaken tambourines, etc.). A precise relationship can now be worked out between the lightning and thunder elements. At the beginning the two can be separated by up to ten seconds, with the interval gradually reduced to one or two seconds at the climax. This will then increase to about ten seconds again towards the end of the piece. The 'rain' elements can be brought in in heavy 'gusts' about a third of the way through, becoming continuous (steady rain) after the climax and gradually fading away somewhere before the end. Several other sound elements could be included if one wished – the wind, trees swaying, bird-song dying away at the beginning and reappearing at the end etc.
2. *Le Mans*
In a motor race all cars start together and gradually become separated. They circle the track time and time again; the sound of each fades in from the distance, reaches a roaring climax and fades away again in the space of a few seconds. Some cars drop out, one or two may crash (emergency sirens etc.); each has to stop occasionally to refuel or be serviced. The sound of a racing car is hard to imitate precisely on an instrument; the children may prefer to use their voices in conjunction with Kazoos or paper and comb. Other instrumental colours could be included to illustrate the passage of time. The Le Mans itself is a 24-hour race and the time of day could be subtly suggested by gradually changing colours and sound patterns.

Bongo, Rust, and *Gemstones* are less specific in their implied structures. A relatively indeterminate sound picture could be generated for each. Bongo could no doubt be given an imaginary programme, a lion hunting and killing its prey for instance, but a precise structure might best be avoided. Purely abstract, musical considerations can decide the length and frequency of a given type of material. I will shortly be describing at some length a piece of a similarly descriptive nature. This will, I hope, give some further indication of how to build up a more general kind of piece.

Finally, *Jack of Diamonds* could be patterned more or less specifically on a given card game. In chapter 2 I suggested the use of playing cards to determine rhythms. This procedure could easily be adopted here, with a pack dealt and 'played' exactly in the manner of a game. The form the piece might take could be as indeterminate as the game itself. It must however follow the 'rules', and these must now be decided.

Stage 4
The final stage involves the practical working out of all the accumulated ideas. The vast majority of tiny details should now be decided in rehearsal. If possible tapes should be made so that the children can criticize the result of a given trial and make suggestions for its improvement. The precise organization of the piece can be decided in various ways:
1. It can be left entirely to the judgement of the individual players when they should start and when they should stop, how long they, as individuals, should play for, how frequently, etc. etc. This takes a great deal of organization, especially with a large class.
2. The piece might have a conductor, who gives signals according to some prearranged plan. This plan can be as general or as specific as one wishes and the conductor can use either the 'improvisation signs' of Chapter 2, or a number system (raised fingers etc.), or any other method which the pupils decide.

3. Alternatively a general outline may be put up on the blackboard. This can range from the simplest diagram to a detailed score. Each pupil can on the other hand jot down an outline on his part to have in front of him and so on. Any number of different methods could be used including a mixture of any of those suggested above.

To conclude this chapter I would like to describe in some detail the making of a simple piece. It was composed by a group of second-year pupils and subsequently performed by them at a school concert. The central idea of the piece was 'radiated energy' and the title chosen was – *Heat.*

Contrary to what one might have expected, it was decided to make a very quiet and sustained interpretation of this idea. The emphasis was to be on the relentlessness of radiation, in other words the kind of heat given by an electric fire, or the sun on a cloudless summer day. All associations with 'fire', 'burning', 'sweltering' etc. were discounted. A central sound was chosen as a pivot around which to build; a central sound which corresponded with the central idea. This was a high sustained cluster on the electronic organ with a fairly pronounced vibrato. This cluster was to be played continuously throughout the piece and could only be varied very slightly in its internal constitution. All other sounds were to be subordinated to the central sound and the aim was always to either match or blend in with the organ. Instrumental colours which could 'imitate' the central sound most closely were given the longest total durations and the more a sound deviated from the central sound, the less it was allowed to appear. One or two sounds were in fact only added occasionally at points of highest density.

Here is a complete list of all the sounds which were chosen in order of decreasing significance:

1. High organ cluster (the central sound)
2. Rubbed (resonating) glasses (two players)
3. Two other organ clusters at different pitch ranges (these were played by a second organist who also adjusted the colour stops from time to time, but not the octave registration, so that the central sound was able to change gradually from one timbre to another)
4. Two glass bowls making for the most part a continuous tremolo, with the occasional single note repeated slowly and regularly.
5. A group of 4 triangles of different pitches played quietly and randomly.
6. A low xylophone playing continuous trills with occasional improvised figurations very quietly.
7. A set of low chime bars – short glissandi with a hard stick and soft repeated clusters, very quiet indeed.
8. Two marraccas played intermittently as quiet as possible.
9. A group of high glockenspiel notes playing continuous trills.
10. A low ethereal whistle (of human origin) and occasional humming by the players.
11. A harmonica played very softly.

The total effect of the piece was of a continuous slowly changing sound. Everything that happened was felt to be 'inside' the high organ cluster. Very little needed to be precisely planned; the working out process was mainly one of elimination. Several tapes were made until a final 'refined' performance was reached. The result was naturally very static and contemplative, and also very beautiful.

CHAPTER 5

ELECTRONIC MUSIC IN THE CLASSROOM

Although there was a certain foreshadowing in the music of composers such as Messiaen and the early Cage, and even among the 'Dada' composers of 1916 who included Edgar Varèse, electronic music did not begin in earnest until after the second World War. Experiments were first carried out in Paris by Pierre Schaeffer and Pierre Henri. This music, the first to be created on electronic tape, was called 'Musique Concrète'. It was made by distorting live sounds in various ways. Raw materials were sometimes obtained from conventional musical instruments such as the tam-tam, or alternatively taken from recordings of environmental sounds, such as train whistles, dogs barking, people conversing etc. The various techniques for distorting these sounds were a more elaborate version of the techniques which are later to be described in this chapter; the speeding up and slowing down of sounds, playing the sounds backwards, cutting the tape in various ways to produce unusual effects, and so on. All these were used to produce patchwork collages of distorted natural sound.

The school of electronic music which grew up shortly afterwards in Germany with such composers as Stockhausen, Pousseur, and Eimert, used as the basis of their compositions, only 'synthetic' material or sounds which are actually created electronically. These basic sounds included sine waves (the purest of all possible sounds), white noise (the most complex mixture of frequencies possible), square waves, saw-tooth waves and so on; all of these were synthesized electronically. The school as a whole was opposed to the decorative freedom of 'Musique Concrète' and in particular its use of natural sounds. The structural procedures of the new school were a development of the serialism of Webern and Messiaen and the works they produced were much more disciplined, if rather ascetic.

It was not long however before Stockhausen and Berio, who was working in Italy, introduced at least one natural sound into their electronic compositions: the human voice. Stockhausen's *Gesang der Junglinge* and Berio's *Homage to James Joyce* combined a single human voice with the electronic sounds, distorting it and superimposing it in many ways to produce striking and often highly dramatic effects. Berio's *Differences* takes this process one stage further and uses live sound, distorted in many ways; it makes full use of the stereophonic possibilities of sound reproduction, in a piece which combines live material with its electronically distorted counterpart. In other words certain instruments are heard live and little by little distorted versions of the sounds (from these same instruments) combine with them so that the 'differences' become more and more apparent. More recently still, Stockhausen has produced pieces which use a whole gamut of electronic devices which distort sounds as they occur (in other words there is *no* prerecorded tape and all the sounds are produced live in performance). Because contemporary developments have to a large extent moved away from the pure electronic music of the first German school, I feel justified in applying the expression 'electronic music' even to the very elementary pieces which follow later in this chapter. Technically speaking this music is nearer to the French 'Musique Concrète'.

2

At first it may seem alarming that electronic music should be made in the classroom. The methods outlined in this chapter however require only the most rudimentary techniques and only the simplest equipment. At first one should carry out a large number of experiments and these need only remain rough. If, however, at a later stage one wishes to make a series of 'finished' pieces which might ultimately function as ballets, as incidental music to a play, or as concert pieces with or without 'live' accompaniment, the teacher will find a knowledge of tape splicing an absolute necessity. This is not a difficult process by any means and can be mastered with very little practice. I myself tend to use a very crude method, aligning the two ends of the tape and cutting them diagonally across with a pair of scissors. Many musicians, especially when they have a tape recorder of their own, are used to this simple chore and any further detail would, I am sure, be unnecessary at this stage. In any case most jointing kits will give full instructions as to how to make good joins. I would strongly recommend the teacher to gain some experience of this.

The procedures for making electronic music which are suggested in this chapter involve two basic processes. The first is the process of recording the basic materials, and this can be done in the classroom during the course of a number of lessons. The second process involves the splicing together of the various parts to form entire pieces. The process should involve no more than two or three people. The teacher might take the responsibility for doing the splicing entirely himself or alternatively enlist the services of a small group of pupils, who need not be trained musicians. I am certain that a number of sixth-formers could always be found to take an interest in this aspect of the creative process, pupils who might otherwise take only a very passive interest in music. Some of the material could be recorded by these pupils: in fact the process could be organized and executed by them in its entirety. As a start the teacher could provide the recorded materials himself, and then matters could develop from this point as a series of sixth-form projects.*

3

The twenty pieces of the previous chapter provide the necessary materials for the 'filling in' of the various 'blue prints' which appear in this chapter. To start with a series of experiments should be carried out which will give one an idea of the various types of manipulation possible during the process of recording. The means of producing distortion and the facilities for superimposition will of course vary with the type of tape recorder used. I will

*If the teacher is fortunate enough to have *small* groups of younger pupils, there is no reason why the making of electronic music should be withheld from them. I have mentioned the sixth form largely because there is no longer any problem of control and such a group would tend to be small enough for this activity. Making electronic music along simple lines is not beyond the capacity of even a twelve-year-old.

only deal with the two most standard and straightforward types of tape recorder. Any additional facilities a tape recorder might possess should always be tried and used for experiment even if they are not mentioned here.

The simplest and cheapest tape recorder on the market is the four-track machine with two speeds. (I have a small Stellaphone which has been recommended by the Inner London Education Authority for some time now). This sort of machine tends to be very robust and will allow much switching from one speed to another even as the machine is recording. Another effect which one would never attempt with an expensive machine, is the slowing down of the left-hand spool, again while recording, by applying pressure with the fingers to the outer edge of the spool. Here again a small machine of this kind will suffer no ill effects. This latter process is rather crude and unpredictable in effect, but it nevertheless makes up for the small number of speeds on this machine (usually only 1⅞ and 3¾ i.p.s.). Another feature of four-track machines is that two of the tracks can often be played back simultaneously, which means that two totally different materials, even materials recorded at two different speeds, can be superimposed one on top of the other. On the whole, this sort of tape recorder should be used for producing the most dynamic and experimental types of material. The process of superimposing two tracks should be liberally used when one is experimenting, but because it is generally inaccurate as far as precise juxtaposition is concerned (i.e. one cannot hear the material which lies beneath whilst recording) I have not suggested its use in the eight plans which follow.

A second common type of tape recorder in current use in schools is the two-track model with three possible speeds (1⅞, 3¾, and 7½ i.p.s.). I have a two-track Truvox which is again recommended by the ILEA (there is a four-track model as well). One has generally to be more careful in using this machine not to overwork the speed-change mechanism, and the manipulation of the spools whilst the tape is running is not recommended; it is in any case harder to control. The three speeds on the other hand are a great advantage; the machine will record and reproduce with greater fidelity and, if it is a machine like a Truvox, a striking echo effect can be added which will be described in more detail at the end of this chapter.

For the recording of materials, two basic elements can be controlled; namely the speed of recording and the volume. If a given material is recorded at 1⅞ i.p.s. and played back at 7½ i.p.s. the pitch or relative pitch will have risen two octaves and any basic pulses or rhythms will appear four times as fast. If pressure is applied to the left-hand spool during the recording process, both the speed and the pitch will appear to rise when the material is played back. Careful manipulation of this kind can produce a whole range of glissandi, accelerandi and ritardandi ranging from the most subtle to the most violent.

As far as the contol of volume is concerned, two processes can be applied. The first is simply the manipulation of the volume control whilst recording and the more speedily and extensively the recorded volume is changed, the more striking the effect. The fact that the closer to the microphone a sound is recorded, the louder it appears, can lead to a great deal of variety in the way the various sounds can be balanced against each other. Sounds recorded within inches of the microphone naturally sound very striking particularly if the instrument in question resonates in a particularly subtle way, like a cymbal. If for instance one records a single cymbal stroke, one can make a superb 'electronic' effect by having the volume control on zero for the attack (i.e. at the moment the cymbal is struck) then subsequently turning the volume up and down to catch the resonance. This effect may be expressed graphically as follows:

Many other simple experiments of this kind should be tried. If for instance a large number of instruments are playing loudly and producing a very dense kind of sound material, an oscillating rhythm can be produced by turning the volume control up and down in the following manner:

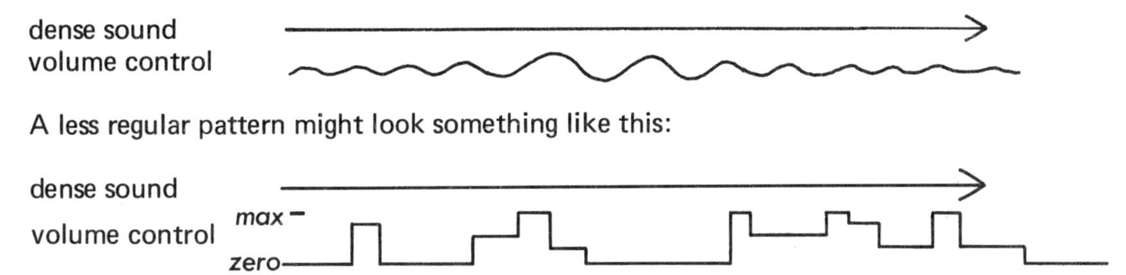

A less regular pattern might look something like this:

This second graph has the effect of cutting the material into blocks of sonority, interdispersed with silence.

Many experiments should be carried out in this manner. The basic material should be produced from the kind of improvisation described in Chapter 2 and one should try and see how many interesting effects one can produce first of all by simply adjusting the volume control. In doing this, the basic types of adjustment described so far should be borne in mind (i.e. the regular wave form patterns of changing volume, irregular patterns involving short stretches of silence, smooth or jagged patterns, etc.). The speeding up or slowing down of a pattern of volume changes is particularly effective. Here is a graphic representation of this process:—

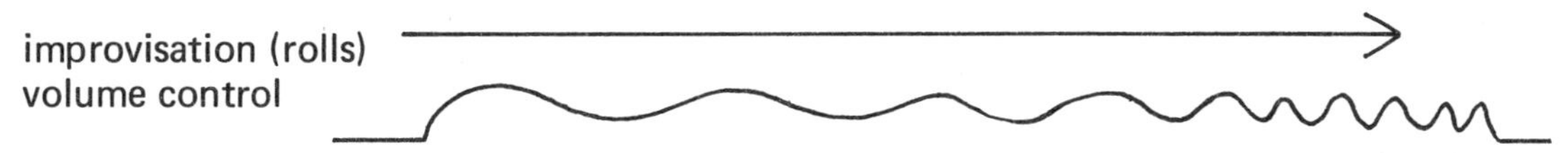

Regarding the use of available speeds, one or two simple experiments should be tried first of all:

First record a given material at $1\frac{7}{8}$ i.p.s. and play back at 3¾ i,p.s., then try this the other way round.

Next extend the process by recording at $1\frac{7}{8}$ i.p.s. and playing back at 7½ i.p.s., finally reversing this process by recording at 7½ i.p.s. and playing back at $1\frac{7}{8}$ i.p.s. With these latter two extremes of speeding up and slowing down one must be careful in the first place to record sufficient material for the final result to be of reasonable length (for example, one needs 40″ of material at $1\frac{7}{8}$ to produce 10″ worth of material at 7½) and if, alternatively, one is slowing the material down considerably, only a very short stretch of material is required to produce something of very long duration.

In all these experiments the type of improvised material should be varied as much as possible and full use should be made of all available colours and effects. One should not only try out single types of texture (i.e. points, rolls, and rhythms taken separately) but also limit the colour to one or two particular sonorities from time to time. In other words an entire range of different types of improvisation should be carried out, each one varied not only with regard to the instruments and textures used but also by using these in conjunction with different recording processes. Pupils should be encouraged to conduct the improvisations when the recording process becomes elaborate so that the teacher's attention can be directed more and more to the tape recorder.

The adjustment of speed whilst recording should now be attempted and this will depend on the type of tape recorder used. Just as interesting patterns can be achieved (regular, irregular, etc.) by turning the volume up and down, corresponding patterns can be made by applying pressure to the left-hand spool while recording. Varying degrees of smooth, violent, regular, and irregular manipulation should be tried.

The previously mentioned process of varying the balance by placing the microphone close to different instruments should now be tried from time to time. One should select an instrument or group of instruments and place the microphone at varying distances from these sources. An enormous variety of different colours can be produced in this way. Finally extensive experiments can be made by combining several of these processes at the same time – the volume being varied while one manipulates the left-hand spool, etc. The teacher should try as many different combinations as possible.

One of the most striking and authentically 'electronic' effects can be obtained by playing a given material backwards. This procedure is especially effective when point sounds are predominent, particularly when resonant instruments are used. Chains of rhythms also sound remarkably effective. The process of attack and decay is reversed so that when an instrument like a cymbal is struck and the result played backwards, the resonance grows out of nothing and appears to crescendo towards the original attack. Treated in this way a piano sounds conspicuously like an electronic organ.

There are two processes by which taped material can be played backwards; that is, if one's tape recorder is not an expensive machine with a simple reversing switch. Of these two methods, the first is not really to be recommended because it results in a considerable loss of quality. It is however the *only* process possible if one has access to one tape recorder only. It consists of simply reversing the *playing side* of the tape and running the material in the opposite direction. (This is not the same as turning the tape over as one does when the end of the spool is reached. This simply selects a new track on the *same* side of the tape. Genuine reversal can only be achieved by either twisting the tape on either side of the playback head, or, preferably, cutting the required material out then joining it back again with the side of the tape reversed *and* the length of the tape in the opposite direction). The smooth side of the tape should then be in contact with the magnetic heads and not the rough side which is usual. Unfortunately, with this process, only half the original volume can be obtained at most and, as I have said already, much of the quality is lost.

The second method produces much better results altogether. Two tape recorders are required, one of which must be a two-track machine and the other a four-track. One can record on either; if one records on the two-track machine, the reverse of the given material can be obtained on track 2 of the four-track machine (i.e. the second track after the spools have been turned over). Alternatively one can record on track 3 of the four-track machine (i.e. the 2nd track of the first side)*. The material should then be played back on track 2 of the two-track tape recorder (i.e. with the tape turned over). These procedures will doubtless become clear when they are tried out experimentally.

These then are some of the simple ways by which material can be electronically distorted. As much experimental material as possible should be obtained using each of these procedures, either separately or together in different combinations. A whole variety of different types of material, live material that is, should be used in conjunction with these methods. Ultimately large quantities of 'electronically distorted' sound textures will have been collected on tape, some of which is bound to be imperfect, uninteresting, inconsistent, or over-distorted. All such waste matter can be discarded. Only the most interesting materials should be retained ready to be fused into larger, more ambitious, 'collages'. Making an electronic piece of this kind is in fact rather like making a film. A very large number of 'takes' have to be made – the film is then finally put together in the cutting room. Just as with film, the cutting process with this sort of music is extremely important and here are two outlines or plans, which can either be utilized in detail or regarded simply as examples of the way this process should be tackled.

Plan No. 1 needs five basic materials of contrasted character. A dash affixed to a number (i.e. 4') implies a variant of the given material, which can refer either to variation in the colour (i.e. instrumentation) or a variation in the way the material has been distorted. Materials 5 and 5' for example, must be recognizably the same despite any deliberate differences. It is assumed that the materials as a whole are improvisatory in character, otherwise choice of materials is entirely free.

To start with, each stretch of material should be cut into segments as follows (1 should be the shortest ingredient and 5 the longest). The total length of the 'varied' materials should be less than the total lengths of the initial materials to which they correspond.

1 1st half	1 2nd half	1' (shorter than 1)		
2 first ¼	2 central ½	2 final ¼	2' 1st half	2' 2nd ½ (shorter than 2)
¾ of 3	last ¼ of 3	1st ¼ of 3'	final ¾ of 3' (shorter than 3)	
¾ of 4	$\frac{1}{8}$ of 4	last $\frac{1}{8}$ of 4		
$\frac{1}{16}$ of 5	¼ of 5	$\frac{3}{16}$ of 5	final ½ of 5	

*Four-track machines can be rather confusing. Here is a simple diagram to show how the tracks are usually labelled

	1st SIDE	2nd SIDE (TAPE TURNED OVER)
POSITION 1	TRACK 1 ———	TRACK 4 ———
POSITION 2	TRACK 3 ———	TRACK 2 ———

N.B. Turning the tape over does not reverse the *playing side* of the tape, despite the fact that one always talks of 'turning the tape over'.

Now follows the plan into which these segments of material should be fitted. Note that the plan as a whole tends to be improvisatory in character, thus providing a free structure for what is a very free kind of material. In other words the structure is designed to match the type of material it will contain.

Improvisatory Plan 1

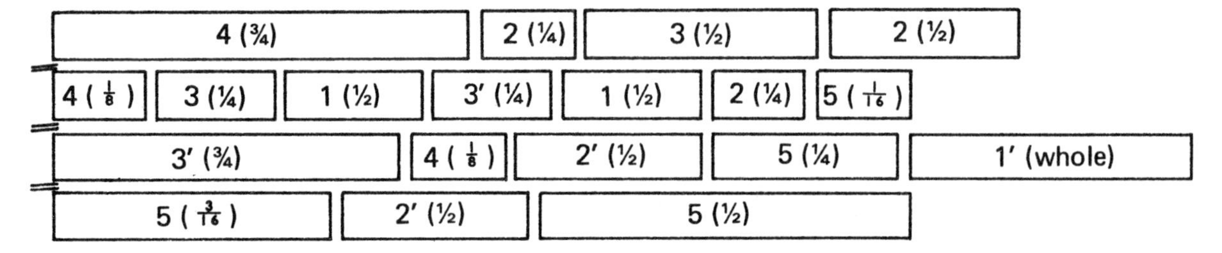

Note the basic film technique of having one or two 'previews' of a given material leading up to a substantial portion of that material (e.g. material 5), and the opposite process of a material gradually diminishing in size and finally disappearing altogether (Materials 3 and 4). Stockhausen's 'Moment' form, with its anticipations and 'memories', is also closely related to this process.

For the second of these two 'open' plans, six contrasted Materials are required which should be divided up as follows (Material 1 is the shortest, 6 the longest as before):

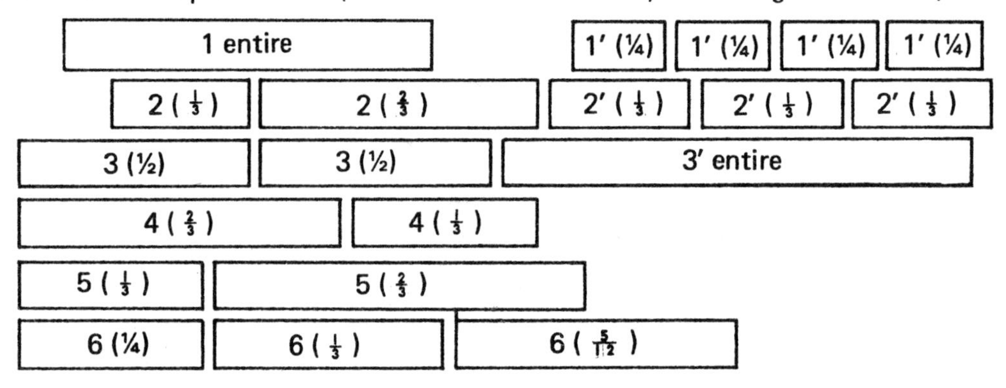

Here is Plan 2 in its entirety. (Exact lengths are of course free as with the first plan. If at any point a pause or short silence is required one should not hesitate to insert one).

Improvisatory Plan 2

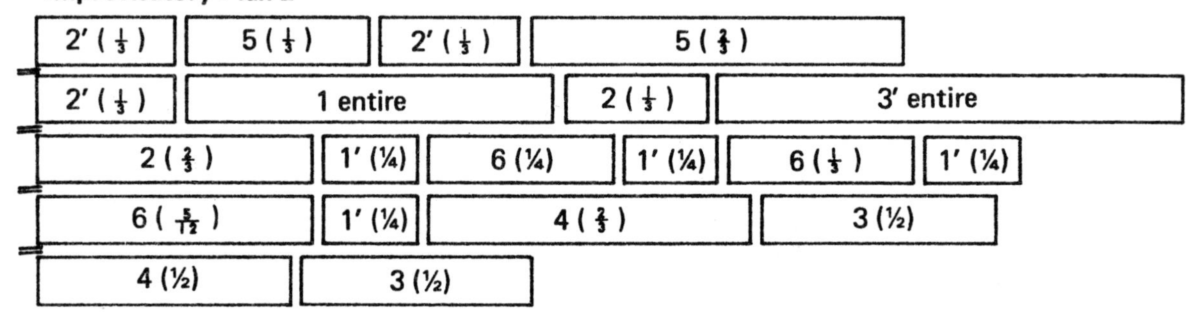

Prominence is given to material 6 which should be of some considerable length. Material 1′ breaks into material 6 in short violent bursts of sound. By way of contrast, Material 6 should be tranquil in character. Another dialogue of 'sounds' takes place between Materials 5 and 2′ but the dynamic contrast need not be so strongly marked here. These are at least a few suggestions as to how this plan can be interpreted which need not be definitely adhered to. The more elaborate specifications of the eight plans set out below should be more closely observed.

The twenty materials of Chapter 3 provide the basic ingredients for these 'blue-prints'. Each is carefully labelled and various instrumentations are suggested. This is not to say that instrumentation will always be elaborately specified. Very often the choice will still rest with whoever is making the piece. What I would like to specify, as a general principle, is that a *limited* number of different instruments should be used for a given piece (i.e. for each entire plan) so that a general consistency of sound is maintained throughout, and this should apply equally to the live accompaniments which *must* 'match' the tape part. Even when up to five or six different materials are used, the instruments chosen for one of the materials (say the one requiring the greatest number of players) should be included as far as possible in the instrumentation of the remainder.

The first two plans in this group are for a small two-speed tape-recorder; plans 3 – 6 require a 3-speed model; plan 7 must have a 2-speed *and* a 3-speed machine and plan 8 requires *two* 3-speed machines. Plans 2, 5 and 8 are pieces for tape alone, which can be used as ballets, incidental music etc.: the remainder are to be performed as concert pieces with live accompaniment.

The following signs are used as shorthand:–

①↑ = *either* record at 1⅞ and play back at 3¾
or record at 3¾ and play back at 7½. whichever is applicable

②↑ = record at 1⅞ and playback at 7½

①↓ = record at 7½ and playback at 3¾

N = record at the same speed as the playback

distorted = careful manipulation of the left-hand spool whilst recording (this applies to the first two plans only)

12′ = a variant of the given material, in this case 12, which *always* refers in this instance to a change of instrumentation.

⟵ = material to be played in reverse *not* electronically but literally, starting the given piece at the end and working backwards.

Plan 1

2 - speed tape-recorder/playback speed 3¾ i.p.s. – performed with live accompaniment

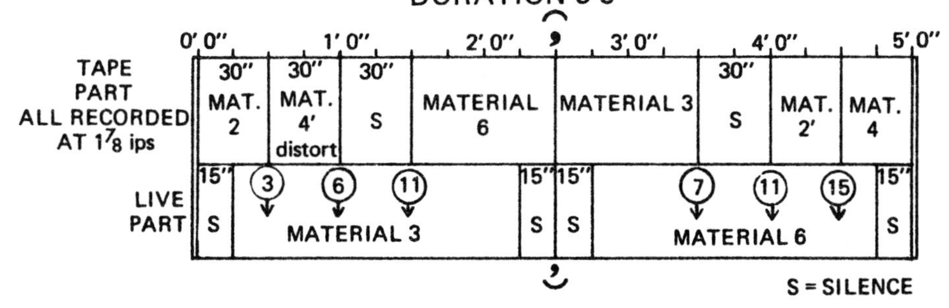

The way in which the live part is played against the prerecorded tape should be fairly obvious from the diagram, although arrows have been inserted in the live parts to give an indication as to how the beats in these materials coincide with the beginnings and ends of the tape materials. A short pause of 2 or 3 secs. may be inserted at the centre of the piece: Material 6 in the tape part should be faded out slowly and then Material 3 gradually faded in after the pause.

Materials 2, 3, 4, and 6 are required for this piece. They should be recorded as follows:
Material 2 should be recorded twice: once normally at $1\frac{7}{8}$ and once distorted at $1\frac{7}{8}$ i.p.s. The time interval between each beat should be reduced to 6'' (for this plan only)
Instrumentation – 2 (normal) – parts 1 & 2 - drums, parts 3 & 4 – metal instruments (with drum-like qualities if possible). For 2'' (distorted) this should be reversed (i.e. parts 1 & 2 – metal instruments and parts 3 & 4 - skin)
Material 3 needs only to be recorded once at $1\frac{7}{8}$
Instrumentation – parts 2, 4, 5, and 6 – same as for material 2 (normal) parts 1 and 3 – wooden instruments (wood blocks, castanets etc.)
Material 4 should be recorded once at $1\frac{7}{8}$ and once distorted at $1\frac{7}{8}$
Instrumentation – parts 'pitched' 1, 2, and 3, should be played by chimebars or glockenspiels.

The remaining parts should be the same as for Material 3 (use any unusual sound for the central roll of the 2nd 'pitched' player) Material 4' is distorted but with the same instrumentation as 4 (changes of beaters can be made if desired).
Material 6 should be recorded once at $1\frac{7}{8}$
Instrumentation – parts 1, 2, and 3 should be played on glockenspiels or chime bars. Parts 4 and 5 should be played on the same *metal* instruments as in Material 2, Part 6 on a gong or suspended cymbal.

The tape of this piece can be made with the materials in the reverse order if desired. The live materials should in this case also be played in the reverse order. The general level of recording should be fairly quiet throughout this piece and this level should also be maintained in the live part.

Plan 2

2 - speed tape recorder/playback speed 3¾ i.p.s. – entirely for tape

DURATION 4'40''

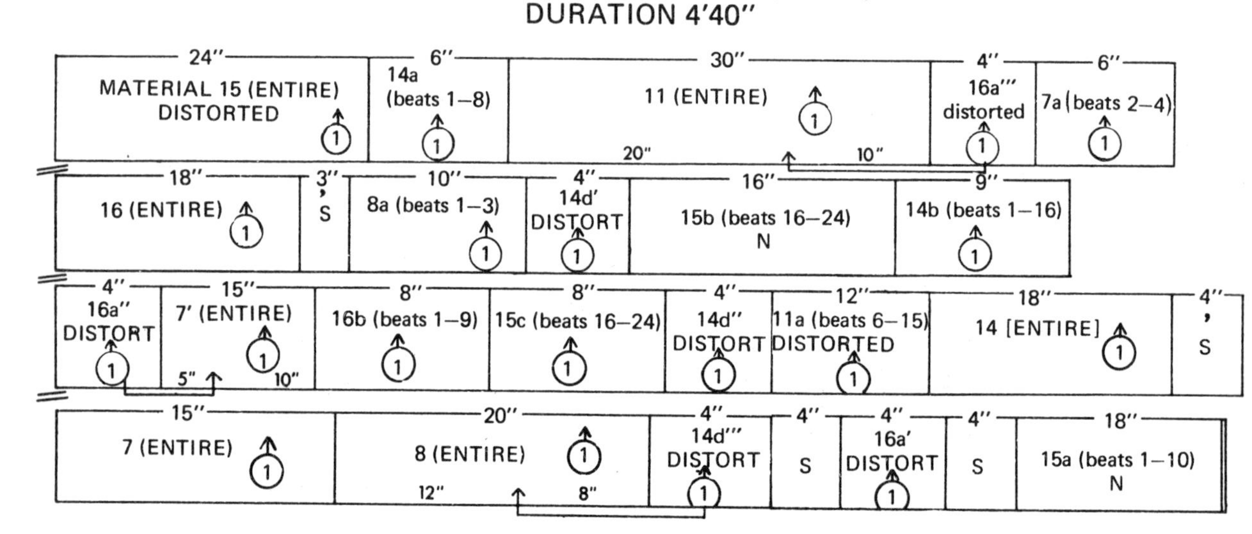

This is the most complex piece in the entire series. The elaborate splicing of the materials is intended to make up for the lack of contrapuntal layering (in other words the fact that the piece contains a linear succession of materials) as well as the tape-recorder only having two speeds.

The six Materials required are 7, 8, 11, 14, 15, and 16. They should be recorded as follows:—

Material 7 should be recorded three times at 1⅞, twice with an even regular beat, and once with a marked accelerando and ritenuto (see the notes for Material 3 in Chapter 3).

Only the material between beats 2 and 4 is required for the second (regular pulse) recording and this should be cut out in readiness.

In the plan, 7 = entire piece (regular pulse)
7′ = entire piece (with accel. and rit.)
7a = beats 2 – 4 only (regular pulse)
Instrumentation — part 1 (one player only) — a wood block for 7 and 7a and a high pitched drum for 7′.

Parts 2, 3, and 4 should be played by metal instruments only (two or three per part). With these, soft sticks should be used for 7 and 7a; hard sticks and/or brushes for 7′.

Material 8 should be recorded twice at 1⅞
8 (in the plan) = entire piece
8a = beats 1 – 3 only (which should be cut out)
Instrumentation — part one should have two players (a high and a low drum) which provide a continuous beat throughout.
Parts 2, 3, and 4 — glockenspiels or chime bars
Part 5 (one player only) — tambourine or some kind of rattle
Part 6 — 2 or 3 medium drums (same instrumentation for 8 and 8a).

Material 11 needs to be recorded twice, once at 1⅞ and once at 3¾
11 (recorded at 1⅞) = entire piece
11a (recorded at 3¾) = beats 6 – 15 only (distorted) (this should be cut out)
Instrumentation —
Part 1 — high drum (one player)
Part 3 - medium drum (one player)
Part 2 — gong, damped each beat (one player)

The three pitched parts should be played on glockenspiels or chime bars; the remaining two parts should be played on (1) high jingles or marraccas and (2) tambourines or rattles. The same instrumentation is used for 11 and 11a.

Material 14 must be recorded three times (at 1⅞) in its entirety but in addition the last beat (i.e. the final cadenza) must be recorded another three times (also at 1⅞)
In the score — 14 = entire piece
14a = beats 1 – 8 (cut out)
14b = beats 1 – 16 (cut out)
14c (three times i.e. 14c′, 14c″, 14c‴) = last beat only
Instrumentation —
Part 2 (one player) — wood block for 14 and 14a; high drum for 14b and 14c. Parts 1, 3, and 4 — metal instruments (instruments already used in previous materials, if possible). Vary the beaters used in parts 1, 3, and 4 for each recording.

Material 15 should be recorded three times altogether. Twice at $1\frac{7}{8}$; and once at $3\frac{3}{4}$.
In the plan, 15 = entire piece at $1\frac{7}{8}$, distorted.
15a = beats 1 – 10 only (recorded at $3\frac{3}{4}$), distorted.
15b = beats 16 – 24, distorted (recorded at $3\frac{3}{4}$, in fact cut from the same recording as 15a)
15c = beats 16 – 24 only (recorded at $1\frac{7}{8}$) not distorted.
Instrumentation –
Parts 1, 2, and 3 – glockenspiels or chime bars. Parts 4, 5, and 6 – drums in 15 and 15c, metal instruments in 15a and 15b
Material 16 should be recorded twice in its entirety at $1\frac{7}{8}$ although beats 1 – 3 need to be recorded another three times on their own (also at $1\frac{7}{8}$)
As indicated – 16 = entire piece
16a (three times i.e. 16a′, 16a″, 16a‴) = beats 1 – 3 only
16b = beats 1 – 9 only (cut out)

The different dynamic levels at which these materials and portions of materials should be recorded can be varied a good deal to obtain greater contrast. The choice of levels is left free although I would suggest that the shortest portions of Materials 14 and 16 (the six Materials 14c and 16a) should be very loud indeed. The arrows which lead on three occasions from these groups into larger adjacent groups, indicate an alternative positioning whereby each of the short groups can be inserted as a violent interruption into the material indicated. If it is felt that this process is too complicated, these materials can be positioned as they occur between groups.

As I have already said, Plan 2 is the most complex in the series. The timing however does not have to be as exact as in some of the other plans, because only one single layer is involved. Nevertheless with its insertions of short 'silences' (which can be lengthened or shortened at will) and its use of a large number of materials cut up in a variety of different ways, the most elaborate process of cutting and jointing is required. Plan 2 should in fact only be attempted after a good deal of experience has already been acquired.

The general character of Plan 2 should be one of strong contrasts. Some of the Materials should however appear to lead on from each other without a break. Material 16 acts as the reverse of Material 14 and should have very similar instrumentation. Material 15 at one stage seems to combine the characteristics of 14 and 16; Materials 7 and 8 also relate to these textures. Because this plan does not have to fit into a rigid time-scheme many liberties can be taken with regard to timing. Silences can be added if felt to be appropriate, and as I have already said existing pauses can be narrowed, extended, or even dispensed with altogether.

Plan 3
for three-speed tape recorder
playback speed = $7\frac{1}{2}$ i.p.s
to be played with live accompaniment — DURATION 5′22″

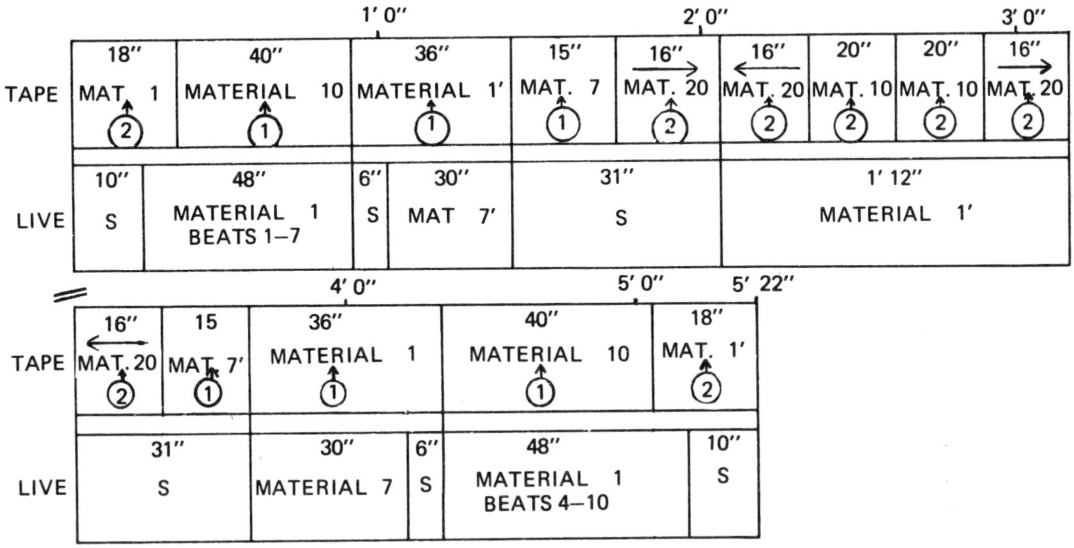

1′ 0″ — 2′ 0″ — 3′ 0″

TAPE	18″ MAT. 1 ②	40″ MATERIAL 10 ①	36″ MATERIAL 1′ ①	15″ MAT. 7 ①	16″ → MAT. 20 ②	16″ ← MAT. 20 ②	20″ MAT. 10 ②	20″ MAT. 10 ②	16″ → MAT. 20 ②

LIVE	10″ S	48″ MATERIAL 1 BEATS 1–7	6″ S	30″ MAT 7′	31″ S	1′ 12″ MATERIAL 1′

4′ 0″ — 5′ 0″ — 5′ 22″

TAPE	16″ ← MAT. 20 ②	15 MAT. 7′ ①	36″ MATERIAL 1 ①	40″ MATERIAL 10 ①	18″ MAT. 1′ ②

LIVE	31″ S	30″ MATERIAL 7	6″ S	48″ MATERIAL 1 BEATS 4–10	10″ S

Materials 1, 7, 10, and 20 are needed for Plan 3. They should be recorded as follows:
Material 1 should be recorded twice at $1\frac{7}{8}$, the second time as a variant of the first (1′), and twice more at 3¾, the second time again being a varient of the first.

Instrumentation

Material 1 – all wooden instruments

Material 1′ – all metal instruments

Material 7 needs to be recorded twice altogether at 3¾, the second time as a variant (7′). The pulse should be completely regular for both.

Instrumentation

Part 1 – one metal plus one skin instrument (same for both recordings)

Parts 2, 3, and 4 – all wooden instruments for Material 7

all metal instruments for Material 7′

Material 10 – should be recorded twice at $1\frac{7}{8}$, the second time as a variant (10′) and twice more at 3¾, the second time again as a variant

Instrumentation

Material 10 – parts 1, 2, and 3 – wooden instruments

parts 4 and 5 – metal (low)

Material 10′ – parts 1, 2, and 3 – metal instruments

parts 4 and 5 – low skin instruments

(The relative pitch markings in the score should always be maintained).

Material 20 is recorded four times altogether at $1\frac{7}{8}$. The arrow ⟵ means that the entire piece should be played in the reverse order – starting at beat 32 and working back to beat 1. This applies to two of the recordings.

Instrumentation – parts 1, 2, and 3 – xylophones or wooden bars (if possible). Parts 4, 5,

and 6 – metal instruments changing to wood half way through (i.e. each player has two instruments), this order is of course reversed when the material is played backwards.

Parts 7 and 8 – low drums

With some of the Materials, notably 10, it is best to fade the material in and out gradually by means of the volume control, while recording. The crescendos in Material 1 should be exaggerated to produce the best effect. This piece (or plan) as a whole deals with a series of waveform repetitions which should be as regular and even as possible (e.g. Materials 10 and 20).

Note that Material 1, in the live part, appears in two incomplete portions and once, at the very centre of the piece, in its entirety. The portions consist of beats 1 – 7 and beats 4 – 10 respectively. The conductor should however gauge the exact positioning of the live materials from the progress of the tape part. Due to the nature of the materials, synchronization does not have to be absolutely exact.

Plan 4

three-speed tape recorder

playback speed = 3¾ i.p.s.

to be played with live accompaniment DURATION 5'50"

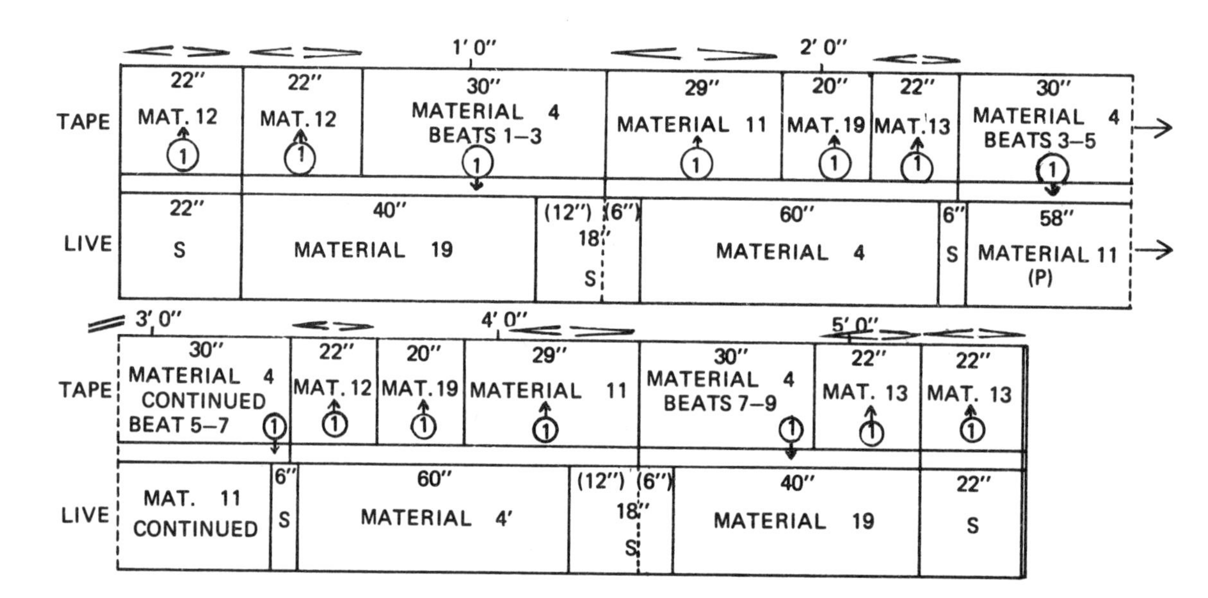

Five materials are needed for this plan; these are 4, 11, 12, 13 and 19. Here is how they should be recorded:

Material 4 needs only to be recorded once at 7½. It should then be cut into portions as follows:—
1st quarter (beats 1 – 3); central section (beats 3 – 7);
final quarter (beats 7 – 9)
Instrumentation –
Parts 2, 7, and 9 (numbering from the top) should be played by metal instruments. The three pitched parts by glockenspiels or chime bars. Parts 1, 3, and 8 should be played on glass instruments. (Players of Part 5, i.e. the second pitched part, should use unusual metal instruments for their central roll.)

A variety of different beaters should be used by the players. In the variant of Material 4 which occurs in the live part, changes should be made both with regard to the beaters and in the way the instruments are played (use different striking areas, etc.). Players of part 5 should have different auxiliary instruments on this occasion.

Material 11 should be recorded twice at 1⅞ and both recordings should be identical.
Instrumentation
The top three parts should have one player each – the instruments should be low triangle, metal drum (damped) and medium low drum. The pitched parts should be played on chime bars.

The lower two parts should be played on marraccas and sleigh bells respectively.

Material 12 – must be recorded three times at 1⅞ and each recording should be identical.
Instrumentation –
Players 1 – 4 should have one player each and the instruments should be low triangle, glass bowl, metal drum, and medium low drum (skin) Parts 5 and 6 should be played on the marraccas and sleigh bells respectively

Material 13 should be recorded three times at 1⅞ and again each should be identical
Instrumentation –
Parts 1 – 4 should have only one player each and the instruments should be low triangle, glass bowl, metal drum (damped), and medium low drum
Part 5 – suspended cymbal with brushes
Part 6 – suspended cymbal with soft sticks (or gong)

Material 19 must be recorded twice at 1⅞ – each time identical
Instrumentation –
Parts 1 – 3 – glockenspiels or chime bars
Part 4 (one player) – glass bowl   Part 5 (one player) metal drum
Part 7 – sleigh bells   Part 8 – suspended cymbal (brushes)

Plan 4 is designed to be tranquil and aesthetic in effect. The live part should always be generally quiet, especially the central material of the piece (material 11). Materials 11, 12, and 13 should be faded in and out as they appear in the tape part. The effect of these materials must be very regular as far as pulse is concerned. The accents which appear in 12 and 13 can be omitted, if desired, to make for a smoother effect.

Plan 5
three-speed tape recorder
playback speed = 7½ i.p.s.
this piece is for tape only

DURATION 5'18"

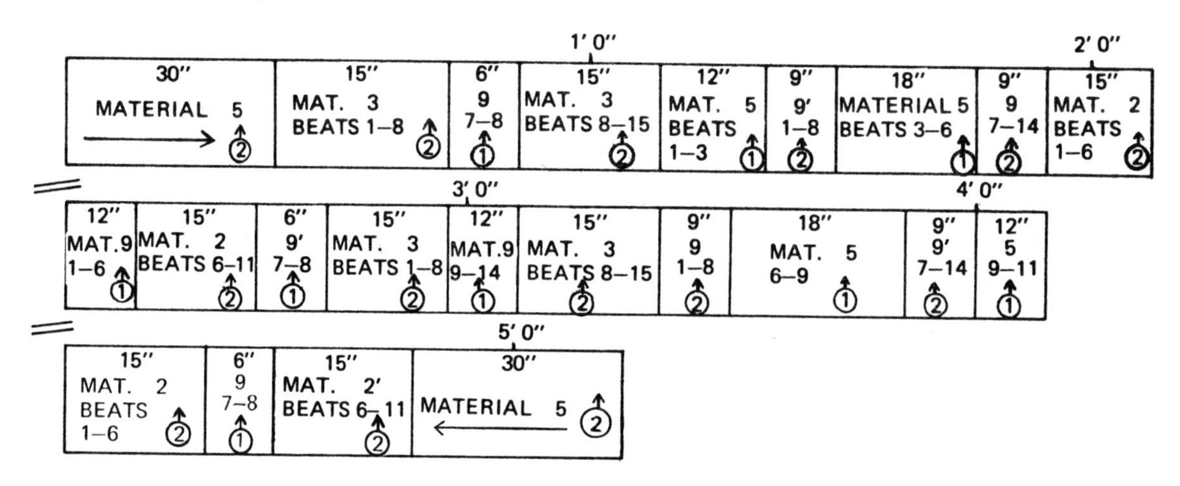

Being something like Plan 2, this piece is above average in complexity; it is a piece which consists of only one layer of sound. Material 9 is the key material of the piece although it never appears complete. The dynamic level of this particular material should be above the general level of the remainder, so that it always functions as a dramatic interruption.

Materials 2, 3, 5, and 9 are used for this plan and should be recorded as follows:

Material 2 needs to be recorded twice at $1\frac{7}{8}$ – the second time as a variant. Each recording should be divided into two equal parts,
i.e. material 2 – beats 1 – 6 and beats 6 – 11
material 2' – beats 1 – 6 and beats 6 – 11

Instrumentation
Material 2 – parts 1 & 2 – metal; parts 3 & 4 – skin
Material 2' – parts 1 & 2 – skin; parts 3 & 4 – metal
(Different metal and skin instruments should be used for each)

Material 3 needs to be recorded twice at $1\frac{7}{8}$. The second recording should be a variant of the first. Each recording should be cut into two equal parts (just as with material 2).
i.e. Material 3 – beats 1 – 8 and beats 8 – 15
Material 3' – beats 1 – 8 and beats 8 – 15

Instrumentation

Material 3		Material 3'	
	part 1 – bottles		part 1 – wood block
	part 2 – high drum		part 2 – high cow bell
	part 3 – small cymbal		part 3 – med. high drum
	part 4 – med. drum		part 4 – low cow bell
	part 5 – suspended cymbal		part 5 – med. low drum
	part 6 – low drum		part 6 – gong

Material 5 should be recorded once at 3¾ and twice at 1⅞. There is no variation in instrumentation although the 2nd recording at 1⅞ should be made with the piece played in reverse (i.e. starting with beat 11) The single recording at 3¾ should be cut into four segments as follows:—

bars 1 – 3; bars 3 – 6; bars 6 – 9; bars 9 – 11

Instrumentation

Parts 1, 2 and 3 — glockenspiels (not chime bars)
Parts 4, 5 and 6 — low metal instruments

Material 9

The recording of this material is quite involved. One complete recording only need be made at 3¾ although the central block, beats 7 – 8 must be recorded again separately (also at 3¾).

Four recordings have to be made at 1⅞, two of which must be variants, and various segments have to be cut from these as follows:

Material 9 (beats 1 – 8)
Material 9 (beats 7 – 14)
Material 9′ (beats 1 – 8)
Material 9′ (beats 7 – 14)

Note that all of these contain the central block (beats 7 – 8) which is why four separate recordings have to be made.

Instrumentation

Material 9		Material 9′	
	part 1 — small cymbal		part 1 — high drum
	part 2 — med. high drum		part 2 — high cow bell
	part 3 — med. low drum		part 3 — low cow bell
	part 4 — gong		part 4 — low drum

Plan 5 is generally quite a dramatic piece and recording levels should all be reasonably high. Material 5 is the only material which should be at all tranquil and, for best effect, should be faded in when it occurs at the beginning of the piece and faded out at the end (the end of the entire plan that is; it should *not* be faded at the end of its first appearance but lead directly into material 3).

Plan 6

three-speed tape recorder
play back speed = 3¾ i.p.s.
all the material is reversed electronically on tape: the piece should be played with live accompaniment

DURATION 5′20″

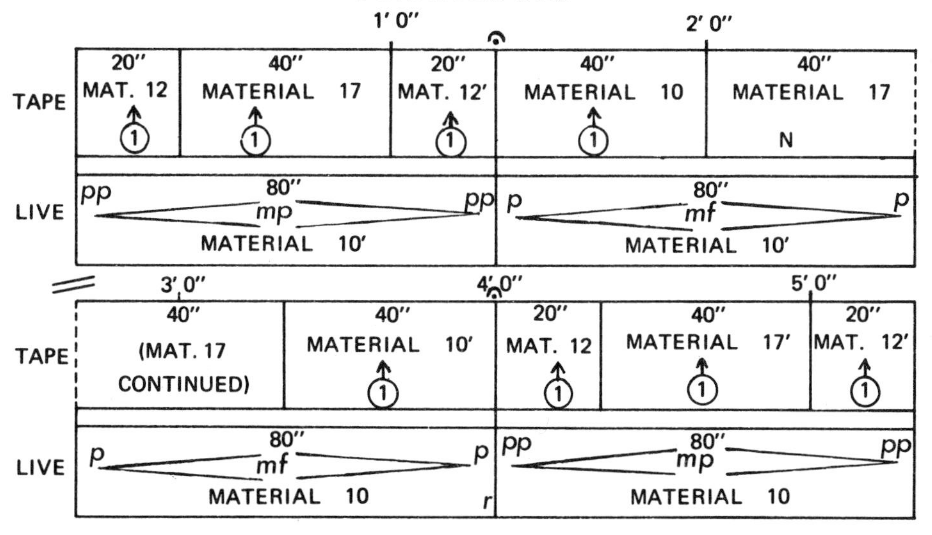

The materials of this plan should be played backwards on the tape. This should be done in the manner indicated earlier in this chapter (the second method if possible). All recordings can be done on a four-speed tape recorder (on the 2nd track of side 1, i.e. track 3) and the result played back on a two-track machine (on side 2). Apart from this the plan is relatively easy to construct. The following three materials are needed — 10, 12 and 17. They should be recorded as follows:

Material 10 should be recorded twice at $1\frac{7}{8}$ — the 2nd time as a variant.

Instrumentation —

Material 10	part 1 — marraccas	material 10′	part 1 — triangle
	part 2 — tambourine (high)		part 2 — sleigh bells
	part 3 — high cymbal (hard sticks)		part 3 — tambourine (low)
	part 4 — large cymbal (brushes)		part 4 — large cymbal (soft sticks)
	part 5 — gong (soft sticks)		part 5 — bass drum

Material 12 has to be recorded four times at $1\frac{7}{8}$. Two of these should be variants. There should be *no* accents (i.e. the *sf* markings in the score should be ignored).

Instrumentation —

Material 12	part 1 — triangle	Material 12′	part 1 — bottles
	part 2 — cow bell		part 2 — small cymbal
	part 3 — cymbal (damped)		part 3 — glass gong
	part 4 — medium drum		part 4 — gong (damped)
	part 5 — sleigh bells		part 5 — marraccas
	part 6 — tambourine		part 6 — large cymbal (brushes)

Material 17 should be recorded twice at $1\frac{7}{8}$ (the second time as a variant) and once at $3\frac{3}{4}$ (i.e. the same speed as playback)

Instrumentation —

Material 17	part 1 bottles	Material 17′	part 1 — cow bell
	part 2 — small cymbal		part 2 — high drum
	part 3 — medium drum		part 3 — tambourine (high)
	part 4 — glass gong		part 4 — tambourine (low)
	part 5 — suspended cymbal (hard sticks)		part 5 — suspended cymbal (hard sticks)
	part 6 — gong		part 6 — bass drum

The general effect of this piece should be tranquil although there are some dynamic clear-cut rhythms in the tape part. Playing the material backwards should produce some quite extraordinary sounds, which should contrast with the deliberately repetitive, almost monotonous, live part. This latter material, which consists entirely of piece 10 played four times altogether, should be very carefully graded dynamically as indicated in the score. This material must also be very consistent rhythmically.

Plan 7

Two tape recorders are needed — one three-speed and one two-speed machine. The materials on the latter are played entirely backwards.

There is also a live accompaniment. DURATION 6′40″

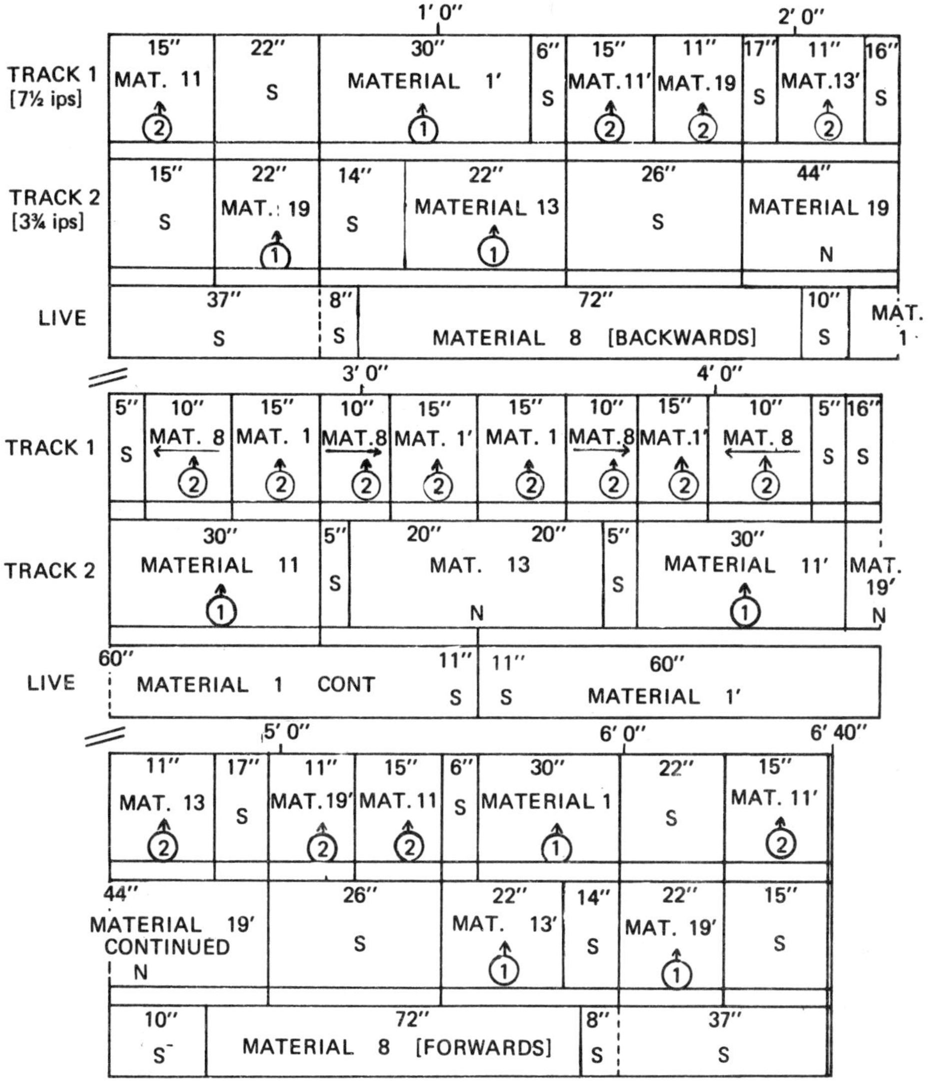

1' 0" — 2' 0"

TRACK 1 [7½ ips]	15" MAT. 11 ②	22" S	30" MATERIAL 1' ①	6" S	15" MAT. 11' ②	11" MAT. 19 ②	17" S	11" MAT. 13' ②	16" S
TRACK 2 [3¾ ips]	15" S	22" MAT. 19 ①	14" S	22" MATERIAL 13 ①	26" S	44" MATERIAL 19 N			
LIVE	37" S	8" S	72" MATERIAL 8 [BACKWARDS]	10" S	MAT. 1				

3' 0" — 4' 0"

TRACK 1	5" S	10" MAT. 8 ← ②	15" MAT. 1 ②	10" MAT. 8 → ②	15" MAT. 1' ②	15" MAT. 1 ②	10" MAT. 8 → ②	15" MAT. 1' ②	10" MAT. 8 ← ②	5" S	16" S
TRACK 2	30" MATERIAL 11 ①	5" S	20" 20" MAT. 13 N	5" S	30" MATERIAL 11' ①	MAT. 19' N					
LIVE	60" MATERIAL 1 CONT	11" S	11" S	60" MATERIAL 1'							

5' 0" — 6' 0" — 6' 40"

TRACK 1	11" MAT. 13 ②	17" S	11" MAT. 19' ②	15" MAT. 11 ②	6" S	30" MATERIAL 1 ①	22" S	15" MAT. 11' ②
TRACK 2	44" MATERIAL 19' CONTINUED N	26" S	22" MAT. 13' ①	14" S	22" MAT. 19' ①	15" S		
LIVE	10" S	72" MATERIAL 8 [FORWARDS]	8" S	37" S				

This is an elaborate plan which requires a great deal of synchronization not only between the tape and the live parts, but also between the two separate tape parts. In performance a number of problems could be solved by having an operator for each of the tape recorders who can start and stop the machines in accordance with the timing when the occasion arises. Alternatively one can measure exact lengths of silent tape which can be inserted in each part in the appropriate places. The live parts are simple and diffuse enough not to matter as far as absolutely precise synchronization is concerned; the tape parts, on the other hand, must be carefully matched and balanced against each other. The following materials are required: 1, 8, 11, 13, and 19.

Each track will be dealt with separately; first track 1 (on the three-speed tape recorder, playing back at $7\frac{1}{2}$ i.p.s.).

Material 1 must be recorded twice for this track at $3\frac{3}{4}$, the second time as a variant.

Instrumentation –

Material 1
part 1 – marraccas
part 2 – sleigh bells
part 3 – medium drum
part 4 – suspended cymbal (soft sticks)

Material 1′
part 1 – triangle
part 2 – castanets
part 3 – suspended cymbal (hard sticks)
part 4 – med. low drum

Material 8 should be recorded four times at $1\frac{7}{8}$ (the entire piece should be played in reverse for two of these, i.e. starting at beat 5 and working back). Note: this material must also be played backwards in the live part.

Instrumentation –

Part 1 – cow bell and medium low drum

The three pitched parts – glockenspiels and xylophones.

The two lower parts – castanets and suspended cymbal (hard sticks) respectively.

Material 11 must be recorded four times at $1\frac{7}{8}$ and two of these should be variants.

Instrumentation –

Material 11
part 1 – high wood block;
part 2 – medium drum
part 3 – suspended cymbal (soft sticks – damped)
parts 4, 5, and 6 – xylophones
part 7 – marraccas
part 8 – tambourine (shaken)

Material 11′
part 1 – cow bell
part 2 – low wood block
part 3 – med. low drum
parts 4, 5, and 6 – glockenspiels
part 7 – castanets
part 8 – suspended cymbal (brushes)

Material 13 should be recorded twice at $1\frac{7}{8}$; the second time as a variant.

Instrumentation –

Material 13
part 1 – triangle
part 2 – low wood block
part 3 – medium drum
part 4 – gong (damped)
part 5 – marraccas
part 6 – suspended cymbal (brushes)

Material 13′
part 1 – high wood block
part 2 – cow bell
part 3 – suspended cymbal (damped)
part 4 – med. low drum
part 5 – tambourine
part 6 – gong

Material 19 is recorded twice at $1\frac{7}{8}$; the second time as a variant

Instrumentation –

Material 19
parts 1, 2 & 3 – glockenspiels
part 4 – med. drum
part 5 – gong (damped)
part 6 – marraccas
part 7 – suspended cymbal (brushes)
part 8 – low drum

Material 19′
part 1, 2 & 3 – xylophones
part 4 – cow bell
part 5 – med. low drum
part 6 – sleigh bells
part 7 – tambourine (shaken)
part 8 – suspended cymbal (soft sticks)

Plan 7 – Track 2 (All of which should be played backwards on the two-track machine)

The instrumentation of the materials is the same as for track 1. All the materials should be recorded on the 3-speed tape recorder and played back on the 2nd track of the 2nd side of the 4-track machine (i.e. track 4). Material 1 and 8 are not required for track 2.

Material 11 should be recorded twice at $1\frac{7}{8}$ (the second time as a variant)

Material 13 should be recorded twice at $1\frac{7}{8}$ (the second time as a variant) and once at 3¾ (the same as the playback speed)

Material 19 should be recorded twice at $1\frac{7}{8}$ (the second time as a variant) and twice at 3¾ (the second time again as a variant)

This piece should, as a whole, be one of fairly strong contrasts. Track 2 is, however, generally quieter than track 1. Balance must of course be carefully maintained between the two tracks during performance. Synchronization should be as exact as possible, as I have already said, particularly at the beginning and end of the piece, when the tracks follow each other without a break. A tiny amount of overlapping is preferable to ominous silences between the materials.

As far as recording levels are concerned, the technique of fading in and out can be used, especially when silence occurs before and after a given material. Sounds growing into prominence and then fading form a kind of counterpoint of overlapping wave formations; the piece as a whole should be shaped in this way.

Plan 8

Two three-speed tape recorders are needed for this plan; they should play back at 7½ i.p.s. and 3¾ i.p.s. respectively

This plan is for tape only. DURATION 6'40"

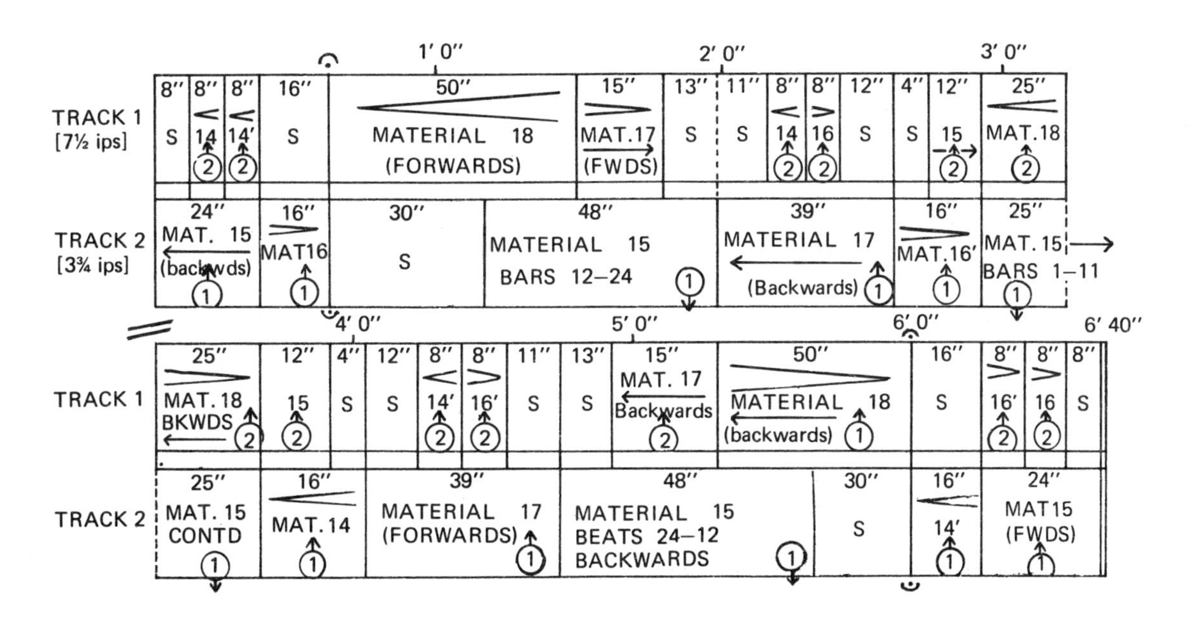

This final plan requires two three-speed tape recorders playing back at two different speeds (7½ i.p.s. and 3¾ i.p.s.). The two tracks must be carefully synchronized with each other as in plan 7.

The following five materials are used: 14, 15, 16, 17, and 18.

Each track will be dealt with separately as before.

Track 1 (tape recorder plays back at 7½ i.p.s.)

Material 14 should be recorded four times altogether at $1\frac{7}{8}$ and two of these recordings should be variants.

Instrumentation –

Material 14	part 1 – sleigh bells part 2 – bottle (one player) part 3 – med. drum part 4 – suspended cymbal (hard sticks)	Material 14′	part 1 – marraccas part 2 – Chinese block (one player) part 3 – glass gong part 4 – suspended cymbal (brushes)

The final beat of material 14 should be played quickly, without diminuendo, and should lead straight into the next material.

Material 15 should be recorded twice at $1\frac{7}{8}$, the second time with the beats in reverse (i.e. beats 24 – 1)

Instrumentation –

Parts 1, 2, and 3 – chime bars (low as possible) – tubular bells or low metallophones.
Part 4 – bottle(s) Part 5 – chinese block Part 6 – med. drum

Material 16 should be recorded four times at $1\frac{7}{8}$ twice as a variant.

Instrumentation –

Material 16	part 1 – bottle (one player) part 2 – castanets part 3 – med. drum part 4 – suspended cymbal (hard sticks)	Material 16′	part 1 –Chinese block part 2 – two bottles (low) part 3 – glass gong part 4 – suspended cymbal (brushes)

Material 17 is recorded twice at $1\frac{7}{8}$ – the second time in reverse (NB. the parts can easily be exchanged to produce a reverse, part 1 and 4 are the same, 2 changes with 6, 3 with 5)

Instrumentation –

Part 1 – Chinese block, bottle and med. drum
Part 2 – marraccas Part 3 – castanets
Part 4 – glass gong Part 5 – suspended cymbal (soft sticks)
Part 6 – deep gong (soft sticks)

Material 18 should be recorded twice at $1\frac{7}{8}$ (the second time in reverse) and twice more at 3¾ (the second time again in reverse)

Instrumentation –

Part 1 – two bottles for trill (one only for last section)
Part 2 – marraccas Part 3 – sleigh bells
Part 4 – glass gong Part 5 – Chinese block
Part 6 – med. drum

(NB. one player only for each part)

Track 2 (playback speed – 3¾)
The instrumentation is the same as for track 1.
Material 14 should be recorded twice at 1⅞, the second time as a variant
Material 15 should be recorded twice at 1⅞ (the second time in reverse). The entire piece must also be recorded twice at 7½, the second time again in reverse. The following segments should be cut from these two recordings:
Bars 1 – 11 and bars 12 – 24 – from the 'forwards' version and bars 24 – 12 only, from the 'backwards' version (this segment should last 50" exactly).
Material 16 should be recorded twice at 1⅞, the second time as a variant.
Material 17 needs to be recorded twice at 1⅞, the second time with the material in reverse. (Material 18 is not used in Track 2)
Two pauses are indicated in the plan; one near the beginning and the other near the end. These should be of two or three seconds each. Materials 14, 16, and 18 should always be recorded, on both tracks, with marked crescendos or diminuendos which ever is appropriate (these are indicated in the score). As in Plan 7, care must be taken not only to make sure that the two tracks are accurately synchronized, but also that balance is maintained between them as far as dynamics are concerned. The general character of the piece involves sudden rhythmic interruptions against relatively tranquil material. With the materials which have to be played in reverse, it is quite permissible to have them played backwards electronically rather than physically. This process would however add considerably to the complexity of making this piece, seeing that all the remaining material on the tracks has to be played forward in the conventional way. If it is felt that this added complication is worth the trouble, it can be applied not only to this plan but to several previous plans which require the reverse of certain materials. One is at liberty to use either procedure.

SUMMARY

With the majority of the 'plans' in this chapter, the splicing together of the various materials is a painstaking and often highly intricate process.The construction of the tape should never be rushed. As I have already suggested the entire process could be undertaken by a few willing senior pupils, and it could be highly beneficial to them as an intellectual and artistic exercise. A quick way of labelling each material must be found (a small label attached to the beginning of each stretch of tape by means of jointing tape would be best) as well as somewhere to hang long lengths of material. Exact timing of materials is more necessary in some pieces than in others, but a stop watch is an indispensable piece of equipment, and one at least should always be available. Most other practical problems can be solved with a little ingenuity.

When these pieces are performed, extra amplification should always be used in conjunction with the tape recorders, unless the performance is taking place in very confined surroundings. With two tape recorders as in the two final plans, stereophonic separation of sound is preferable, but this does require two amplifiers, unless of course one is lucky enough to have access to a stereophonic mixer. Even with only one tape recorder two loud-speakers are preferable, and in the pieces with live accompaniment these should be placed on either side of the performing ensemble.

The pieces for tape alone can, as I have already said, be used for a variety of purposes – ballets, eurhythmics, incidental music to plays, and so on. Best of all perhaps, such a piece could be played in conjunction with a series of changing visual images, making use of any number of 'visual aid' devices – slide projectors, episcopes etc. It might even be possible to produce moving film to correspond to the 'sound images' of the various pieces, so that a collage of visual effects becomes part of the overall experience.

Whereas the 'Eight Plans' are generally smooth-flowing and repetitive in character, the Improvisatory pieces outlined earlier in this chapter can be highly 'expressionistic' in character. Materials of this kind can range from the most dramatic or grotesque to the most serene and delicate. A really well made 'improvised' collage, however spontaneously it may have been made, can be just as valid a piece of 'sound music'.

There is one final way by which material can be very effectively distorted and I have deliberately left it to the end of the chapter because of its wider implications. This process needs a tape recorder, such as the Truvox recommended earlier, which can add an electronic echo. This is done very simply, during recording, by turning up the 'playback' volume control. This type of machine has three magnetic heads, the erasing, the recording, and the playback heads, which are set out in that order from left to right. This means that one can record and play material back simultaneously – a given sound appearing again after a short delay. This delay depends on the speed at which one is recording: and because the microphone picks up the playback sound from the speaker instantaneously, an echo is produced which has the effect of repeating the initial sound over and over again. The slower the speed the slower the echo and the less the sound becomes distorted. At 1⅞, the echo produced is quite slow (just over half a second).

The higher the playback volume is raised, the louder the echo becomes. Feedback will then accumulate and distort the initial sound. Used sparingly this particular type of feedback is not unpleasant, and can be used to produce an interesting 'electronic' side effect. If the playback volume is kept just under the volume of the original sound, the echo will gradually die away. With improvised materials, this effect should be used from time to time alongside other forms of manipulation. It is also very effective if material produced in this way is played backwards.

Echo is however only one of several methods, most of which are too complex and expensive for use in school, by which live sound can be transformed into 'electronic' sound at the moment of playing. One should try improvising a series of point sounds with added echo, tuning the playback volume up and down at will. The tape recorder must of course be on 'record' (1⅞ and 3¾ produce the best results) and a tape must be running through. Nevertheless the effect produced is a spontaneous result of the sounds which are fed into the microphone. The positioning of the microphone is of course very important and the closer it is to a given sound source, the more striking the effect.

This then is an additional form of electronic manipulation which can be used not only to produce interesting improvised materials for a tape collage, but also as a means of adding variety to classroom improvisations. This, among the many other forms of improvisation, manipulation, and composition outlined in this book, can help teacher and pupil alike to discover a new world of sound; a world which has much in common with our own environment and with the sounds and technology of our age.